INDUSTRIAL STEAM LOCOMOTIVES

Displaced by dieselisation, three saddle tanks await a decision on their future at Desford Colliery, Leicestershire, in 1966. All supplied by Peckett and Sons Bristol: 1175, built 1908; 1733, built 1927; and 1587, built 1924.

INDUSTRIAL STEAM LOCOMOTIVES

V. F. Hall

BRACKEN BOOKS
LONDON

First published in 1977 by Moorland Publishing Company

This edition published in 1988 by Bracken Books, an imprint of Bestseller Publications Ltd., Princess House, 50 Eastcastle Street, London W1N 7AP, England.

ISBN 1 85170 238 5

Printed in Great Britain by
The Camelot Press, Southampton

Foreword

By the time this book is in print the steam locomotive in industry will be virtually extinct. Perhaps it was because it lacked the glamour of the main-line locomotives that it was ignored by the majority of railway enthusiasts and even sometimes looked on with scorn by them.

In compiling this book I have tried to select a good cross-section of locomotives produced by the major British manufacturers and while not purporting to be a comprehensive album, the illustrations selected show many of the standard designs found in industrial service. Some early designs, or those of smaller manufacturers, are not illustrated due to either lack of photographs in the author's collection or because those available were unsuitable for reproduction. Similarly those locomotives which had been so rebuilt as to be not truly representative of the manufacturer have also been omitted. Nevertheless it is hoped this book will convey to the reader some of the character of industrial locomotives and the atmosphere attached to the railway systems on which they worked.

For those who wish to see the prototype locomotives, or who may wish to model a particular design illustrated, an appendix of preserved industrial locomotives has been included. It is not a complete list, but covers many of the preservation societies. However, locomotives in private preservation have been omitted in fairness to their owners.

The Industrial Steam Locomotive

It is often forgotten that the first steam locomotives were built for industrial use. The early development of railways can be traced back to the plateways and wagonways laid down by industry to facilitate the movement of wagons pulled by horse or man power. Although Richard Trevithick, the Cornish inventor, is credited with building the first locomotive in 1804, it was not until later that the industrialists realised its potential. In 1822 George Stephenson built locomotives for use on the Hetton Colliery Railway in County Durham. Traditionally coal had been conveyed in barges from the upper reaches of the Wear and trans-shipped into seagoing vessels at Sunderland Docks. This operation was necessarily costly and it was realised that a direct railway line from the pits to Sunderland would enable them to undercut their competitors. The foresight in authorising the building of these locomotives gave the necessary impetus to ensure the development of the steam engine. Other colliery owners followed suit, not only in North-East England, but also in Lancashire where similar problems were encountered. Soon it was realised that there was a market for passenger as well as freight traffic and so the main-line railway company evolved.

In the early days private companies were encouraged to provide not only their own wagons, but in some cases the locomotives also, which had running rights over the main lines. Some of these rights were exercised well into British Railway days; the ex-Lambton, Hetton and Joicey system being an example.

As railways took over from canals and the packhorse, so the demand for locomotives grew. This was the era of railway mania when schemes were manifold and investors stood to make or lose a fortune, depending on the viability of the project. It was the zenith of the railway contractor never seen before on such a scale and unlikely to happen again.

During the years 1850-70 Britain was in a boom period. Industry was expanding putting greater demands on raw materials, and new quarries and mines were being opened up to meet this. It was not now necessary to locate industry in traditional areas, but instead it could be sited near to a source of raw materials, skilled labour, or where a necessary climatic or geological feature was present, and the missing requisites brought in by railway. As factories and works became larger, so it was necessary to have a means of internal movement of wagons within the premises, or to transport materials between the works and the exchange sidings or canal wharf. Mine and quarry companies used industrial railways to convey material from the working face to the processing plant, while locomotives were used extensively by contractors for major civil engineering works such as docks and reservoirs. For instance, it is recorded that 173 locomotives were used in the construction of the Manchester Ship Canal.

It was for these markets that the industrial locomotive builders designed their products. At first similar designs were produced for both industrial and main line use, but as the loads and running speeds on the main lines increased, so the locomotive designs were enlarged to cater for this. There was thus a need for a slow-speed shunting engine having a short wheelbase and a light axle-loading. Initially, these were produced by rebuilding redundant main-line locomotives,and the legendary Isaac Watt Boulton founded a thriving business in Ashton under Lyne based on that practice. The first true industrial designs appeared by the mid-1850s when E. B. Wilson and George England produced locomotives for the industrial market. Other firms followed suit and soon even moderate-sized engineering firms with suitable equipment were building locomotives to meet the growing demands of industry. The most common industrial design was the tank engine having four or six coupled wheels with outside cylinders for the former, and a choice of inside or outside cylinders for the latter. The water was generally carried in a saddle tank over the boiler, although side tank designs were quite common and some operators, such as the Manchester Ship Canal Company, even standardised on them. Well tanks and tender engines were also supplied to industry, though in much smaller numbers.

As often happens, a slump followed the boom period resulting in mergers and liquidations, so that by the turn of the century locomotive building in Great Britain had been consolidated into eight specific areas. Scotland boasted three areas. Kilmarnock was the home of Andrew Barclay, Son & Company, Limited, who had been building locomotives since 1859. A near neighbour, Barclays and Company, started business in the early 1870s, but in 1888 the two companies merged and continued trading under Andrew Barclay's name. This company latterly produced both steam and diesel shunting-locomotives and is still in existence, having been acquired by Hunslet Holdings Ltd. in recent years. A third firm was Grant Ritchie and Company whose designs were similar to their Kilmarnock rivals. Airdrie also had a thriving railway industry, though the total number of products was much less than the output from Kilmarnock and the firms ceased trading much earlier. The firms established here were Gibb and Hogg, Dick & Stevenson and Inglis & Company. Finally there were the three Glasgow giants, Neilson, Dubs and Sharp Stewart which had moved from Manchester in 1888. In 1903 these three merged to form the North British Locomotive Company which continued manufacturing main line and industrial locomotives, with emphasis on Neilson designs for the latter. After an unsuccessful attempt to establish itself as a manufacturer of diesel locomotives, North British closed down in the early 1960s.

Traditionally North-East England was the home of locomotive building and it is not surprising that several successful firms were established there. Robert Stephenson and Hawthorn Leslie and Company were located at Newcastle on Tyne, though after 1900 Stephensons moved to Darlington and built both main line and industrial locomotives. With declining fortunes the two companies merged in 1938 to form Robert Stephenson and Hawthorns Ltd., which later became a part of English Electric. Black Hawthorn and Company was another builder based at Gateshead, who took over from Coulthard and built a large number of locomotives over a period of several years.

Manchester latterly boasted two companies: Nasmyth Wilson and Beyer Peacock and Company. The latter established a worldwide reputation for reliable main line and industrial locomotives and will be best remembered for its Garratt locomotives. It belatedly entered the diesel field, but failed to gain a foothold and closed down.

The West Riding of Yorkshire had without doubt the greatest number of firms with collectively the largest output. Leeds was the principal city with no fewer than six firms producing locomotives. The smallest firm was T. Green & Son, which built mainly narrow-gauge locomotives for local markets in the early twentieth century. Manning Wardle, on the other hand, started in 1859 and took over the interests of E. B. Wilson producing their designs. Their products were popular with contractors and they enjoyed worldwide success. However, they were unable to withstand the depression of the 1920s and closed down in 1926. Their patterns were acquired by Kitson and Company, another pioneer manufacturer, and for a while Manning Wardle designs were available as an alternative to Kitson's traditional engines. This firm also ceased trading in the 1930s when the joint patterns and designs were acquired by Robert Stephenson and Hawthorns. In 1860 Hudswell Clarke & Rogers was formed, later to become Hudswell Clarke & Company, who built almost exclusively for industrial and light-railway markets. They also foresaw the opportunity for internal combustion traction and diversified sufficiently to survive the depression. The company finally closed in the 1960s. The Hunslet Engine Company was established in Leeds in 1864 and found a ready market for its products with contractors and the many Yorkshire colliery companies. An early producer of petrol and diesel locomotives, Hunslet is still in business today, and as recently as 1971 built a new steam locomotive for export to Indonesia. The last Leeds company was John Fowler & Company. They had established a reputation for traction engines and road rollers but later entered the locomotive field. Their steam products were mainly for the narrow-gauge lines, but having turned over to internal-combustion locomotives they won a large share of the standard-gauge market too.

Sheffield supported only the Yorkshire Engine Company, which later became a part of the United Steel Company. They concentrated on industrial designs and naturally supplied large numbers to steel companies. Although they were successful with diesel traction, they closed down in the 1960s.

The Midlands had three major companies located quite independently of each other and, strangely enough, away from the traditional engineering area of the Black Country. At Loughborough was to be found the Brush Company. This firm had taken over from the Falcon Engine Company and quite early in this century they ceased production of steam locomotives and turned over to diesel traction. They are now part of the Hawker Siddeley Group producing for main-line companies. W. G. Bagnall established his firm in Stafford in 1875 and had moderate success with a multitude of designs. In 1891 however the company was joined by E. E. Bagguley from Hawthorn Leslie, who introduced a range of locomotives, not unnaturally similar to Hawthorn Leslie's own designs, which was to ensure the company's success. Although successful with diesel-locomotive production Bagnall ceased business in the 1960s. The most recent firm to enter the steam locomotive field was Kerr Stuart and Company whose works were founded in Stoke on Trent in 1892. They produced numerous narrow-gauge locomotives, as well as some designs for standard gauge railways. They were early producers of petrol locomotives, but were unable to survive the depression, and closed down in 1929.

The final area was in Bristol, where two companies were to be found. Fox Walker and Company built nearly four hundred locomotives between 1864 and 1880, when falling orders made it impossible to continue trading. In that year Thomas Peckett established his Atlas Works and used modified Fox Walker designs for his basic range of industrial locomotives. Peckett and Sons produced locomotives solely for industrial and light railway markets, where they gained a reputation for reliability and robustness. The company achieved a high degree of standardisation and boasted a fast spares service. However, they were late competitors in the diesel field, which they entered in 1956, and with declining orders they finally ceased business in 1961.

Edwin Walker of Fox Walker founded a new company in Bristol taking the name of Avonside Engine Company. Their range of locomotives at first resembled contemporary Fox Walker and Peckett products, but later a more austere design was introduced. Avonside was another victim of the depression and closed down in 1933.

As the locomotives built by the main line companies were instantly recognisable, so each industrial manufacturer incorporated certain features in his design that became the hallmark of his products. The style of worksplate fitted to the locomotive was peculiar to each manufacturer. Generally these were oval, but the following shapes were also used at some time: triangular (Dubs), diamond (Nasmyth Wilson), rectangular (Grant Ritchie), and round (North British). During its lifetime, however, a locomotive would undergo some rebuilding, which, depending on the nature of the work, would alter its appearance. Many early Peckett locomotives for instance, when reboilered in later years, acquired the small dome and 'pop' safety valves in place of the spring-balance variety originally fitted, while more austere saddle-tanks replaced exotic designs on other manufacturers' products. Cabs, or lack of, also attracted attention and could either be modified in a somewhat Heath Robinson manner, or be properly rebuilt by the manufacturer or operator. Smaller companies tended to deal with locomotive builders or specialist firms for their repairs, whereas many larger operators had workshop facilities that would do justice to a main-line railway company. Here locomotives were rebuilt from side tank to saddle tank or from tender to tank engine with almost monotonous frequency; the Philadelphia works of the Lambton, Hetton and Joicey Colliery Company being an example. Some companies even built their own locomotives, often from parts supplied by a manufacturer; the steel companies were leaders in this.

Steam power did not remain unchallenged for many years however. Electric traction was introduced into industrial use early in the twentieth century, while petrol locomotives were common on the narrow-gauge systems from the 1920s, brought about by the surplus War Department motive power built for World War I. The depression of the 1930s was a difficult time for the industry and Bagnalls are reputed to have built locomotives at a loss in order to maintain work. Other companies were not so fortunate and were compelled to close down. World War II made great demands on industry and many locomotive builders benefitted from the orders for the Ministry of Supply and other military establishments. This was to have repercussions later when the locomotives became surplus and were purchased by industrial operators, thus depriving the manufacturers of potential orders for new machines.

By the end of the 1950s the days of the steam locomotive were numbered. Diesel traction was being produced in increasing numbers by a variety of manufacturers and proving its superiority over the steam locomotive to both operator and driver. Parallel to this was the development of heavy self-propelled plant and machinery for civil engineering work and the use of shunting tractors and materials-handling equipment, which eliminated the need for rail transport altogether. In quarries the high-capacity tipping lorry and conveyor belt displaced the narrow-gauge railway, while standard-gauge systems felt the effects of the Beeching plan which closed many lines and private sidings. Other recent developments that have enforced the closure of many industrial locations are the change to natural gas, increased use of oil pipelines and the trend on British Rail towards block trains with automatic loading and unloading.

Today the National Coal Board and British Steel Corporation are the major users of industrial locomotives, though they too have felt the effects of rationalisation. Fortunately many industrial steam locomotives survive on preserved railway lines where they represent an important part of the history of steam-locomotive development.

The illustrations that follow are arranged by manufacturer, in alphabetical order, with four-coupled locomotives generally preceding their six-wheeled counterparts and narrow gauge appearing before standard gauge. Unless stated otherwise, locomotives are to standard gauge, i.e. 4ft. 8½in. The final group of photographs, commencing with Fig. 106, illustrate industrial locomotives at work at their location.

Abbreviations

LOCOMOTIVE TYPE

T	Side tank
CT	Crane tank
PT	Pannier tank
ST	Saddle tank
WT	Well tank
VB	Vertical boilered locomotive
F	Fireless locomotive
G	Geared locomotive
IC	Inside cylinders
OC	Outside cylinders
VC	Vertical cylinders

WHEEL ARRANGEMENT

The Whyte system of designation is used throughout. However, where wheels are chain driven as in Sentinel locomotives, the wheel arrangement is shown thus:4W.

LOCOMOTIVE MANUFACTURERS

AB	Andrew Barclay Sons & Co. Ltd., Kilmarnock
AE	Avonside Engine Co. Ltd., Bristol
AP*	Aveling Porter Ltd., Rochester
BH	Black Hawthorn & Co. Ltd., Gateshead
BP	Beyer Peacock & Co. Ltd., Manchester
D	Dubs & Co. Ltd., Glasgow
EB	E. Borrows & Sons, St. Helens
FW*	Fox Walker & Co., Bristol
GH	Gibb & Hogg Ltd., Airdrie
GR	Grant Ritchie & Co., Kilmarnock
HC	Hudswell Clarke & Co. Ltd., Leeds
HE	Hunslet Engine Co. Ltd., Leeds
HF*	Richard Evans, Haydock Foundry
H(L)*	Hawthorns, Leith
HL	R. & W. Hawthorn Leslie & Co. Ltd., Newcastle upon Tyne
K	Kitson & Co. Ltd., Leeds
KS	Kerr Stuart & Co. Ltd., Stoke on Trent
Mkm	Markham & Co. Ltd., Chesterfield
MW	Manning Wardle & Co. Ltd., Leeds
N	Neilson & Co., Glasgow
NBL	North British Locomotive Co. Ltd., Glasgow
NR	Neilson Reid & Co., Glasgow
NW	Nasmyth Wilson & Co., Manchester
P	Peckett & Sons Ltd., Bristol
RS	Robert Stephenson & Co., Newcastle upon Tyne
RSH	Robert Stephenson & Hawthorns Ltd., Darlington and Newcastle upon Tyne
S	Sentinel Ltd., Shrewsbury
SL	Stephen Lewin, Poole
VF*	Vulcan Foundry Ltd., Newton Le Willows
WB	W. G. Bagnall Ltd., Stafford
YE	Yorkshire Engine Co. Ltd., Sheffield

OPERATORS

APCM	Associated Portland Cement Manufacturers
CEGB	Central Electricity Generating Board
MOPWB	Ministry of Public Works and Buildings
NCB	National Coal Board

An asterisk indicates that although examples from this manufacturer are included in the list of preserved locomotives (pp 93-6) similar examples are not illustrated.

Industrial Locomotives Illustrated

Fig	maker	number	type	location
1	AE	1386	0-4-OST	Alders Paper Mills, Tamworth, Staffordshire
2	AE	1591	0-4-OST	NCB Birch Coppice Colliery, Warwickshire
3	AE	1723	0-4-OST	John Baker & Bessemer, Kilnhurst, Yorkshire
4	AE	1547	0-6-OT	Stanton Ironworks Company, Leicestershire
5	AE	1567	0-6-OST	Stanton Iron Company, Ilkeston, Derbyshire
6	AE	2048	0-6-OST	NCB Arley Colliery, Warwickshire
7	WB	2135	0-4-OST	BICC, Erith, Kent
8	WB	1537	0-4-OST	East Midlands Gas Board, Grimesthorpe Works, Sheffield
9	WB	2108	0-4-OST	Bass Worthington & Company, Burton on Trent, Staffordshire
10	WB	2907	0-4-OST	Staveley Iron Company, Derbyshire
11	WB	2223	0-6-OST	NCB Manvers Main Colliery, Yorkshire
12	WB	2192	0-6-2T	Bowater Lloyd, Sittingbourne, Kent
13	WB	2668	0-6-OST	Cranford Ironstone Company, Northamptonshire
14	WB	3061	0-6-OST	NCB Cadley Hill Colliery, Derbyshire
15	AB	1994	0-4-OWT	Penrhyn Quarries, Caernarvonshire
16	AB	1871	0-4-OWT	Scottish Gas Board, Granton Works, Edinburgh
17	AB	2030	0-4-OCT	Stanton Iron Company, Ilkeston, Derbyshire
18	AB	2046	0-4-OST	British Aluminium Company, Burntisland, Fife
19	AB	2323	0-4-OST	South Durham Iron Company, Irchester Quarries, Northamptonshire
20	AB	1000	0-6-OST	NCB Gedling Colliery, Nottinghamshire
21	AB	2141	0-6-OST	Stanton Iron Company, Ilkeston, Derbyshire
22	AB	2025	0-6-OST	NCB Manvers Main Colliery, Yorkshire
23	AB	2273	0-6-OT	Stanton Iron Company, Ilkeston, Derbyshire
24	AB	2193	0-6-OF	CEGB Carlisle, Cumberland
25	AB	2068	0-6-OT	Wemyss Private Railway, Methil, Fife
26	BP	6841	0-4-4-OT	NCB Baddesley Colliery, Warwickshire
27	BH	266	0-4-OST	Bowne and Shaw, Wirksworth, Derbyshire
28	BH	912	0-4-OST	Scottish Gas Board, Aberdeen Works
29	BH		0-6-OST	Staveley Iron Company, Derbyshire
30	EB		0-4-OWT	United Glass and Bottle Company, Charlton
31	D	4101	0-4-OCT	Shelton Iron & Steel Company, Etruria, Staffordshire
32	GH		0-4-OST	NCB Cardowan Colliery, Lanarkshire
33	GR	272	0-4-OST	NCB Fordell Colliery, Fife
34	GR	805	0-4-OST	NCB Cardowan Colliery, Lanarkshire
35	GR	536	0-4-2ST	NCB Lady Victoria Colliery, Midlothian
36	HL	2800	0-4-OST	CEGB Bow, Essex
37	HL	2657	0-4-OST	NCB Bentinck Colliery, Kirkby in Ashfield, Nottinghamshire
38	HL	3780	0-4-OST	South Durham Iron Company, Irchester Quarries, Northamptonshire
39	HL	3825	0-6-OST	Stewarts and Lloyds, Corby, Northamptonshire
40	HL	3899	0-6-OST	NCB Upton Colliery, Yorkshire
41	HL	3026	0-6-OT	NCB Wellesley Colliery, Methil, Fife

42	HL	3904	0-6-OT	CEGB Hams Hall, Coleshill, Warwickshire
43	HC	1338	0-4-OST	NCB Elsecar Colliery, Yorkshire
44	HC	1696	0-4-OST	Oxfordshire Ironstone Company, Banbury, Oxfordshire
45	HC	1891	0-4-OST	NCB Skiers Spring Colliery, Yorkshire
46	HC	822	0-6-OST	NCB Manvers Main Colliery, Yorkshire
47	HC	1308	0-6-OST	Stewarts and Lloyds, Corby, Northamptonshire
48	HC	607	0-6-OST	Stewarts and Lloyds, Corby, Northamptonshire
49	HC	1419	0-6-OT	Oxfordshire Ironstone Company, Banbury, Oxfordshire
50	HC	1877	0-6-OT	NCB Brookhill Colliery, Pinxton, Nottinghamshire
51	HC	859	0-6-OT	NCB Wellesley Colliery, Methil, Fife
52	HC	940	0-6-OT	Manchester Ship Canal Company, Irlam, Lancashire
53	HE	317	0-4-OST	Penrhyn Quarries, Caernarvonshire
54	HE	1403	0-4-OST	John Knowles Ltd, Woodville, Derbyshire
55	HE	1493	0-4-OST	Shackerstone Railway Society, Leicestershire
56	HE	1557	0-4-OT	NCB Walkden, Lancashire
57	HE	1446	0-6-OST	Stewarts and Lloyds, Corby, Northamptonshire
58	HE	498	0-6-OST	NCB Haunchwood Colliery, Warwickshire
59	HE	3872	0-6-OST	Oxfordshire Ironstone Company, Banbury, Oxfordshire
60	HE	2882	0-6-OST	NCB Coppice Colliery, Ilkeston, Derbyshire
61	HE	1475	0-6-OT	NCB Astley Green Colliery, Lancashire
62	KS	1049	0-4-2ST	Bowater Lloyd, Sittingbourne, Kent
63	KS	3123	0-4-OST	NCB Walkden, Lancashire
64	KS	4043	0-6-2T	Bowater Lloyd, Sittingbourne, Kent
65	KS	3068	0-6-OT	NCB Astley Green Colliery, Lancashire
66	K	5158	0-6-OST	NCB Walsall Wood Colliery, Staffordshire
67	K	5358	0-6-OT	NCB Chase Terrace, Staffordshire
68	K	2509	0-6-OPT	NCB Derwenthaugh, Durham
69	K	4263	0-6-2T	NCB Philadelphia, Durham
70	SL		0-4-OST	Seaham Harbour Company, Durham
71	MW	1877	0-6-2T	Bowater Lloyd, Sittingbourne, Kent
72	MW	865	0-6-OST	MOPWB Kidbrooke
73	MW	1795	0-4-OST	South Durham Iron Company, Irchester Quarries, Northamptonshire
74	MW	1601	0-6-OST	MOPWB Kidbrooke
75	MW	1762	0-6-OST	Stewarts and Lloyds, Corby, Northamptonshire
76	Mkm		0-4-OST	George Slater Ltd, Beighton, Derbyshire
77	NW	454	0-4-OST	Brown Bayley and Company, Sheffield
78	N	3629	0-4-OST	Bairds and Scottish Steel Company, Coatbridge, Lanarkshire
79	NR	5760	0-4-OST	Bass Worthington and Company Ltd, Burton on Trent, Staffordshire
80	NBL	18385	0-4-OST	Bairds and Scottish Steel Company, Coatbridge, Lanarkshire
81	NBL	16463	0-6-OST	NCB Wellesley Colliery, Fife
82	NBL	24564	0-6-OT	NCB Haunchwood Colliery, Warwickshire
83	P	1258	0-4-OST	South Durham Iron Company, Irchester Quarries, Northamptonshire
84	P	1823	0-4-OST	Ambrose Shardlow and Company, Sheffield
85	P	1880	0-4-OST	Reeds Paper Mills, Greenhithe, Kent
86	P	2119	0-4-OST	Stewarts and Lloyds, Bromford Bridge, Warwickshire
87	P	2029	0-6-OST	Stewarts and Lloyds, Wellingborough, Northamptonshire
88	P	1378	0-6-OST	Alpha Cement Company, Shipton on Cherwell, Oxfordshire

89	P	1787	0-6-OST	NCB Newdigate Colliery, Warwickshire
90	P	1745	0-6-OST	NCB Arley Colliery, Warwickshire
91	S	6515	4WTG	Thos. E. Gray, Burton Latimer, Northamptonshire
92	S	9369	4WTG	Thos. E. Gray, Burton Latimer, Northamptonshire
93	S	9552	4WTG	NCB Manvers Main Colliery, Yorkshire
94	RS	3801	0-6-2T	NCB Philadelphia, Durham
95	RSH	7818	0-4-OST	CEGB Castle Donnington Leicestershire
96	RSH	6947	0-6-OST	NCB Gedling Colliery, Nottinghamshire
97	RSH	7688	0-6-OST	NCB Philadelphia, Durham
98	RSH	7298	0-6-OST	NCB Cadley Hill Colliery, Derbyshire
99	RSH	7668	0-6-OST	Stewarts and Lloyds, Corby, Northamptonshire
100	RSH	7765	0-6-OT	NCB Ashington, Northumberland
101	RSH	7609	0-6-OT	NCB Ashington, Northumberland
102	YE	2345	0-4-OST	English Steel Company, Woodhouse Mill, Derbyshire
103	YE	327	0-6-OST	Stewarts and Lloyds, Corby, Northamptonshire
104	YE	2197	0-6-OST	NCB Bentinck Colliery, Nottinghamshire
105	YE	2413	0-6-OST	Stanton Iron Company, Ilkeston, Derbyshire
106	AB	705	0-4-OST	NCB Pye Hill Colliery, Derbyshire
107	P	1891	0-6-OST	NCB Manvers Main Colliery, Yorkshire
108	RSH	7049	0-4-OST	NCB William Pit, Cumberland
109	{AB AB	1338 1785	0-6-OT	NCB Waterside, Ayrshire
110	AB	2335	0-6-OT	NCB Waterside, Ayrshire
111	AB	2135	0-4-OST	Stewarts and Lloyds, Corby, Northamptonshire
112	HC	431	0-6-OST	Sheepbridge Iron Company, Desborough Warren, Northamptonshire
113	AB	1047	0-4-OST	South Durham Iron Company, Storefield Quarries, Northamptonshire
114	P	1289	0-4-OST	South Durham Iron Company, Storefield Quarries, Northamptonshire
115	K	5477	0-6-OST	Stewarts and Lloyds, Glendon, East Northamptonshire
116	P	1871	0-6-OST	Stewarts and Lloyds, Wellingborough, Northamptonshire
117	P	1867	0-6-OST	Oxfordshire Ironstone Company, Banbury, Northamptonshire
118	RSH	7846	0-6-OT	CEGB Hams Hall, Coleshill, Warwickshire
119	P	1990	0-4-OST	CEGB Ironbridge, Shropshire
120	HC	1730	0-4-OST	CEGB Mexborough, Yorkshire
121	WB	2178	0-4-OST	Alpha Cement Company, Shipton on Cherwell, Oxfordshire
122	P	2112	0-4-OST	J.C. Staton, Tutbury, Staffordshire
123	RSH	7405	0-4-OST	APCM Swanscombe, Kent
124	{P P	1890 1938	0-6-OST	Ford Motor Company Dagenham, Essex
125	HL	2491	0-4-OST	British Celanese, Spondon, Derbyshire
126	P	1376	0-4-OST	British Aluminium Company, Burntisland, Fife
127	HC	1334	0-6-OT	Oxfordshire Ironstone Company, Banbury, Oxfordshire

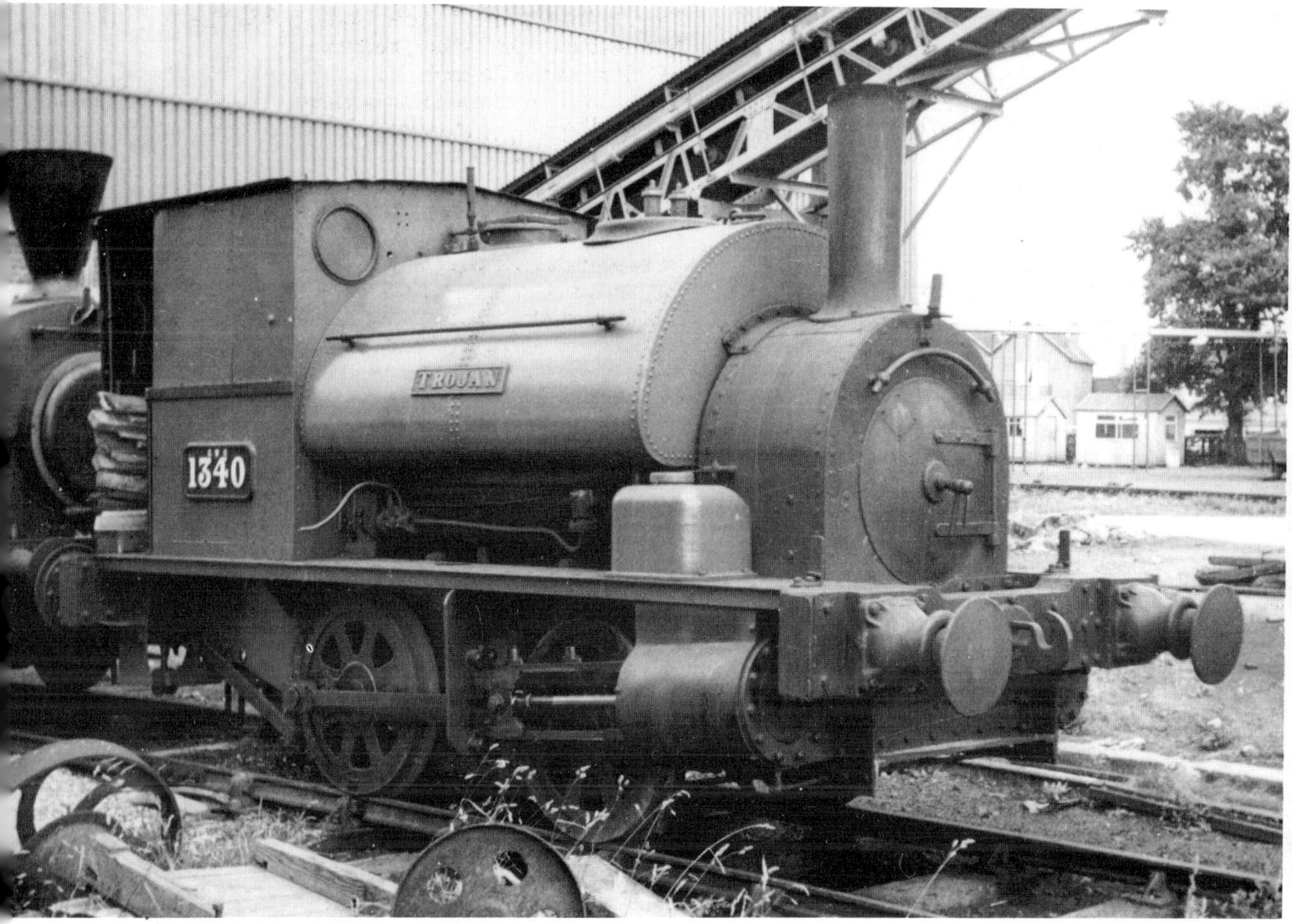

1 Avonside Engine Company, 1386, built 1897

Trojan illustrates the early Avonside design of 0-4-OST in which the round saddle-tank extends from the cab front to the rear of the smokebox. It originally had a bell-shaped dome cover with spring balance safety valves, which were replaced on reboilering in 1922. It was delivered new to Messrs. Dunn and Shute for work at Newport Docks, but in 1903 was acquired by Alexandria (Newport and South Wales) Docks and Railway Company, eventually passing into Great Western Railway ownership. *Trojan* was sold back into industrial service in 1932 and was subsequently bought in 1946 by Alders Paper Mills Ltd., Tamworth, Staffordshire where it was photographed in 1965.

2 Avonside Engine Company, 1591, built 1911

A later design from Avonsides Works is shown to advantage by *Empress.* The round section saddle-tank now extends full length from cab front to smokebox front, while the typical Avonside dome cover incorporates the safety valves. Not surprisingly *Empress* bears a considerable resemblance to contemporary products from that other Bristol manufacturer, Peckett & Sons. Delivered new to Morris & Shaw Ltd., Birch Coppice Colliery, Warwickshire, *Empress* survived well into NCB ownership and is seen there out of use in 1965.

3 Avonside Engine Company, 1723, built 1915

One of the surviving steam locomotives at John Baker and Bessemers Steelworks at Kilnhurst, Yorkshire, was Avonside number 1723. This was to that manufacturer's final design introduced during World War I, which incorporated a new angular design of saddle tank giving the locomotive a sturdy and impressive look. The safety valves are mounted on a squat dome, while the flared chimney has a copper cap. The cab side cut-out is of standard Avonside design. Supplied new to John Baker and Bessemer Ltd in 1915, it was photographed at Kilnhurst in 1964, by which time the works had been closed down.

5 (opposite, bottom) Avonside Engine Company, 1567, built 1909

Stanton Ironworks Derbyshire was the home of a number of Avonside Engine Company products similar to *Stanton* No. 3 illustrated. Points of note are the bell shaped dome encasing the safety valves and the round saddle-tank which extends back to the cab but leaves the smokebox exposed. A tapered chimney is mounted on the smokebox, which unusually has outside steampipes. Supplied new to Stanton Iron Company in 1909, *Stanton* No. 3 was rebuilt in 1920 and 1938 and is seen at Stanton Ironworks in 1964.

4 Avonside Engine Company, 1547, built 1908, 3ft. 0in. gauge

The Avonside Engine Company did not build many narrow-gauge locomotives for the home market compared with such manufacturers as Bagnall and Hunslet. *Nancy* depicts one such locomotive which was supplied new to the Stanton Ironworks Company and worked in the ironstone quarries in Leicestershire. It is a pleasing if undistinguished design, with long side-tanks and a full length cab which originally was open backed, but later received a crude back-sheet. A tall chimney and dome complete the design. Now preserved in Ireland, *Nancy* was photographed in Nottinghamshire when in private ownership.

6 Avonside Engine Company, 2048, built 1932

Resplendent in dark-blue livery with red wheels and motion, *Joan* portrays the later design of Avonside Engine Company locomotives. A copper-capped chimney and neat dome incorporating Ramsbottom safety valves is fitted, while the modern cab and flared coal-bunker will also be noted. Altogether an attractive locomotive, whose short wheelbase makes it ideal for colliery work. Delivered new to Arley Colliery Company, *Joan* was photographed at Arley Colliery Warwickshire in 1967.

7 W. G. Bagnall, 2135, built 1925, 3ft. 6in. gauge
What perhaps epitomises the Bagnall narrow-gauge locomotive is the 0-4-OST shown here. This design was produced over a period of sixty years with little alteration in appearance, and in various gauges. Some were fitted with fully-enclosed cabs, others were completely cabless, while many had the type illustrated. Note the inside frames found even on 2ft. gauge locomotives, and the Bagnall Price valve gear. *Sir Tom* is pictured at BICC, Erith, in 1966, where it was used on their internal railway system.

8 W. G. Bagnall, 1537, built 1898
Typifying the older Bagnall locomotive, *No 1* was photographed out of use at Sheffield's Grimesthorpe Gas Works in 1966. The round saddle-tank covers only the boiler barrel, while the Ramsbottom-pattern safety valves are located over the firebox. Buffers are unusually mounted on separate buffing plates which extend above the top of the buffer beam, while brakes and sanding equipment are operative on both wheels. Also note the commodious cab and the curved handrail fitted on the smokebox front. Delivered new to Grimesthorpe works, *No 1* remained there all its life, apart from rebuilding in 1925 and 1953.

9 W. G. Bagnall, 2108, built 1923

Delivered new to Worthington and Co. Ltd, of Burton on Trent as their No 5, Bagnall 2108 portrays the characteristics of Bagnall's larger 0-4-OST. The saddle tank now covers the smokebox, but again the firebox is left exposed. The small dome cover is a typical Bagnall feature, as are the Ramsbottom-pattern safety valves. On this example a copper-capped chimney is fitted, but the majority sported one of more austere pattern. The locomotive is here seen at Burton on Trent in 1963, by which time Messrs Bass and Worthington had merged and it had been renumbered *16* to avoid duplication of numbers.

10 W. G. Bagnall, 2907, built 1949

To Bagnall and Sons went the distinction of supplying Staveley Iron Company's last steam locomotive. It was a powerful, four-coupled saddle tank with 16in x 24in cylinders and 3ft 6in diameter wheels. The angular saddle-tank was an unusual feature, as Bagnall's tended to favour a round-section tank on their four-wheeled products. *D.N.T.* had a comparatively short working life, being withdrawn in 1965 together with Staveley's remaining steam locomotives. Photographed at Staveley works in 1965.

11 W. G. Bagnall, 2223, built 1924

Elsie is an early six-wheeled product from Bagnall, having a round saddle-tank extending from the smokebox rear to the cab front. The Ramsbottom safety valves are let into the tank over the firebox, while the cab is fitted with the early-pattern flared coal-bunker. The chimney which shows signs of wear was later replaced by one of more conventional design. Supplied new to Tinsley Park Colliery Co Ltd, for use at Kilnhurst Colliery, Yorkshire, *Elsie* was later transferred to Manvers Main Colliery, Wath on Dearne, where it was photographed in 1964.

12 W. G. Bagnall, 2192, built 1922, 2ft 6in gauge

When Edward Lloyd required a large six-coupled locomotive for use on the main line from Ridham Docks to his paper mills at Sittingbourne, Kent, Bagnall & Sons were successful in securing what had hitherto been exclusively a Kerr Stuart market. They produced in total three locomotives with 0-6-2T wheel arrangement, of which *Conqueror* was the first. It had 13in diameter cylinders and 2ft 9in diameter coupled-wheels. Points of interest are the large flat-sided smokebox with balloon-stack chimney, tall, fat dome and safety valves on a small saddle. It is interesting to compare this locomotive with the illustration of *Superior* by Kerr Stuart (Fig 64) to see the resemblance.

14 W. G. Bagnall, 3061, built 1954
Empress is representative of the final design of Bagnall six-wheel saddle tank which was supplied in number to the Staffordshire Collieries and Preston Dock Authority. The points of note are the angular saddle-tank stopping short of the smokebox and typical Bagnall cab and bunker design. It was delivered new to NCB Moira Bath Colliery, Leicestershire, but was later transferred to Cadley Hill Colliery where it was photographed still at work, with two other steam locomotives, in 1974. It is interesting that on arrival at Cadley Hill the locomotive was fitted with the nameplates from the similarly named Avonside product (Fig 2), which had by then been broken up at Birch Coppice Colliery.

13 (opposite) W. G. Bagnall, 2668, built 1942
During World War II Bagnall produced six locomotives for use in the ironstone quarries in Northamptonshire and Oxfordshire. They are identifiable by the full-length round-section saddle-tank, with safety valves on the dome. The cab and coal bunker are standard Bagnall pattern of the day, while the short wheelbase chassis is fitted with outside cylinders. In fact Bagnalls are reputed to have never built an inside cylinder, six-coupled locomotive of their own design. Photographed at Cranford in 1969, *Cranford* No 2 was delivered new to Cranford Ironstone Company, Northamptonshire.

15 (above) Andrew Barclay, 1994, built 1931, 2ft 0in gauge

Glyder was one of a pair of Barclay well tanks operating at Penrhyn Quarries, North Wales, in 1965. The design is unusual for British narrow-gauge practice, in being fitted with Walchaerts valve gear and having the frames located inside the driving wheels. Other features of note are the Belpaire firebox and the fabricated stays supporting the smokebox. Nevertheless it is an attractive locomotive. Supplied new to Durham County Water Board it was acquired by Lord Penrhyn's Slate Quarries in 1938.

16 (opposite, top) Andrew Barclay, 1871, built 1925, 2ft 6in gauge

Many Scottish gasworks utilised narrow-gauge locomotives for internal movement within the works. Typical of Andrew Barclay's products is 1871, which was the last survivor at Granton Gasworks, Edinburgh, where it shared duties with a four-wheel diesel. Points to note are the weather board fitted in place of a cab to reduce overall height, the single slidebars and the manufacturer's worksplate fixed to the cylinder cover. Photograph taken in 1966.

17 (opposite, bottom) Andrew Barclay, 2030, built 1937

The crane-tank design found favour in steel works where it could be used for both shunting and lifting operations. Stanton Ironworks, Derbyshire, had the largest complement of crane tanks in the Midlands, totalling nine locomotives all supplied by Andrew Barclay and Sons. *Stanton* No 33 pictured here is representative of the later designs from that manufacturer The slender jib can only be slewed in the horizontal plane and cannot be raised. It was delivered new to Stanton Iron Company, Ltd, and was photographed at work at Stanton Works in 1964.

18 Andrew Barclay, 2046, built 1937

A highly polished example of a smaller standard-gauge design from this manufacturer. Recognisable features of Barclay's products are the curvaceous full length saddle-tank and large filler set forward of the dome which supports the Ramsbottom-pattern safety valves. With 10in diameter cylinders it is a well proportioned and attractive design. It was the newest of three steam locomotives operated by British Aluminium Company, Burntisland, Fife, in 1965.

19 Andrew Barclay, 2323, built 1952

No 9 contrasts with the previous illustration to show the features that are typical of the later Andrew Barclay designs. The cab has become more angular and the design generally more austere. Ross pop safety-valves are mounted on the dome, while the boiler mountings have been reduced in height to comply with the loading gauge. Supplied new to Dorman Long, Middlesbrough, as their *No 6,* it was later acquired by the South Durham Iron Company's Irchester Quarries, Northamptonshire, and renumbered to avoid duplication by the simple expedient of turning the plate upside down. Photographed at Irchester in 1964.

20 Andrew Barclay, 1000, built 1903

Numerically the 1000th locomotive built by Andrew Barclay and Sons, *Catherine* is a standard 0-6-0ST with outside cylinders, and full-length saddle tank. A tall chimney and cab with flared coal bunker is fitted, but the cab backsheet has been removed. The horizontal cylinders are mounted very low, giving a somewhat squat appearance. It was delivered new to the Digby Colliery Co Ltd, and is pictured at Gedling Colliery, Nottingham, in 1965 shortly before being broken up.

21 Andrew Barclay, 2141, built 1942

An alternative design of six-coupled saddle tank is illustrated by *Stanton No 36*. The saddle tank stops short of the smokebox, which is supported on a separate saddle. The dome is set forward, while the safety valves are let into the tank. Note the ladder from the footplate to the saddle tank to facilitate access to the filler cap. *Stanton No 36* was one of a pair delivered new to Stanton Ironworks Company, where it was photographed in 1964.

22 (above) Andrew Barclay, 2025, built 1936

The latter-day six-coupled design is illustrated by *Tinsley.* A more modern cab is fitted and the flared coal-bunker has been superseded by one with a sharply-angled backplate. A shortened chimney and dome are fitted and overall the design has become more severe. *Tinsley* was supplied new to Tinsley Park Colliery Company, Ltd, and was photographed at Manvers Main Colliery, Wath on Dearne, Yorkshire, in 1964.

23 (opposite top) Andrew Barclay, 2273, built 1949

Stanton No. 38 was the only side tank obtained by Stanton Ironworks Company from Andrew Barclay and Sons and was also the last new steam-locomotive purchased by them. The side tanks have sloping tops to improve forward visibility, while the flattened sides to the cylinders will also be noted. The squat chimney and dome combine with the later cab and bunker design to produce a powerful looking locomotive. Delivered new to Stanton Ironworks in 1949, *Stanton No 38* was photographed there in 1964.

24 (opposite, bottom) Andrew Barclay, 2193, built 1944

The fireless locomotive was used in industries where there was a high fire-risk or pollution from smoke was to be avoided. The boiler was in fact a storage tank into which high pressure steam was pumped from a steam supply in the factory. The steam operated the pistons in the cylinders in the usual way and gave several hours' operation without being refilled. The exhaust chimney is located behind the cab, while the inlet valves for recharging the boiler are at the other end of the locomotive. Carlisle power station had three such locomotives, one of which is shown here in 1965 shunting in the station yard.

W P R
Nº 20

25 (opposite, top) Andrew Barclay, 2068, built 1939

Andrew Barclay did not build many locomotives of their own design with inside cylinders; in fact there were probably more side tanks with this arrangement than saddle-tank locomotives. The Wemyss Private Railway, which was a private company conveying coal from the collieries near Kirkcaldy to Methil Docks in Fifeshire, operated a number of locomotives to the design illustrated. Points of note are the high side-tanks with inspection holes to allow access to the inside motion and the large diameter smokebox topped by a flared chimney. A tall dome is fitted, while the Ramsbottom-pattern safety valves are mounted on a separate base. *No 20* of the WPR fleet is seen in 1965.

26 (opposite, bottom) Beyer Peacock, 6841, built 1937

Beyer Peacock and Company will be best remembered for their Garratt designs, which were supplied to many railway companies throughout the world. Apart from those supplied to the LMS and LNER Companies, Beyer Peacock also built four Garratts for industrial firms in this country. The last survivor was *William Francis,* which was delivered new to Baddesley Collieries, Ltd, Warwickshire, and remained there until withdrawn. It is seen in 1967 at Baddesly Colliery shortly before being sold to Bressingham Museum, Norfolk.

27 (below) Black Hawthorn, 266, built 1873

Holwell No 3 is probably the last industrial locomotive in this country to achieve its centenary in service. It is a typical Black Hawthorn design with a short flat-sided saddle tank, curved weather-board and spring balance safety-valves mounted over the firebox. The chimney is a product of Holwell Ironworks however, replacing the original which would have had a flared top, while the crude cab extension was provided at Wirksworth. *Holwell No 3* was delivered new to contract work, being later acquired by Holwell Iron Company and is pictured in 1970 at Bowne & Shaw's limestone quarry, Wirksworth, Derbyshire, where it was transferred in 1946.

28 Black Hawthorn, 912, built 1887

A later product from Black Hawthorn is seen at Aberdeen where it was employed at the gas works. The locomotive is fitted with a more respectable cab, while the safety valves have moved forward to a centrally-mounted dome. The wheels and motion are unusually fully enclosed, in common with all Aberdeen Gas Works' locomotives, to enable them to work through the streets. *City of Aberdeen* was delivered new to Aberdeen Gas Works, where it was photographed in the early-morning sunlight in 1966.

29 Black Hawthorn, built c1874

Isaac Limb was a six-coupled Black Hawthorn design which was employed at Staveley Ironworks, Derbyshire, for many years. The cab has been rebuilt to afford the driver more protection but otherwise it incorporates all the features shown by *Holywell No 3* (Fig 27). Also note the Ramsbottom-pattern safety valves mounted on the dome. Believed to have been supplied new to a contractor, *Isaac Limb* was later bought by the Staveley Coal and Iron Company and is seen in a scrapyard at Chesterfield in 1961, having been sold by Staveley the previous year.

30 E. Borrows, built 1906

Although the well tank was uncommon on the standard gauge, Edward Borrows & Sons was one firm that standardised on this design. Indeed when Borrows closed down Kerr Stuart acquired their patterns and produced several more identical locomotives. As Borrows works was located in St Helens, it was perhaps natural that their products should have become popular with the glass companies of that town. Recognisable features are the lack of side or saddle tank and the side cab-bunkers accommodating the coal. The flared side to the smokebox is distinctive, while the Ramsbottom safety valves on a separate saddle will also be noted. *The King* is seen in 1966 at the United Glass and Bottle Company's Charlton Works, London, to where it was transferred from St Helens in 1923.

31 Dubs, 4101, built 1901

For many years the last surviving industrial example of this manufacturer could be found at Shelton Iron & Steel Works, Staffordshire, where the crane tank illustrated was used until recently. The design is not noteworthy, but has one or two interesting features. The substantial plate-steel jib moves only in the horizontal plane, while the boiler has the safety valves mounted immediately behind the smokebox so as to clear the crane. The single slide-bars will be noted, as will the solid dumb-buffers which is a feature often favoured by steelworks. Delivered new to Shelton, Dubs 4101 is photographed there in 1967.

32 Gibb & Hogg, built 1898

Gibb and Hogg were Airdrie-based locomotive builders whose designs generally followed the pattern of the Kilmarnock manufacturers, but with certain detail differences. On *No 14* illustrated can be seen the typical pattern of saddle tank, which is a more curvaceous design than those found on Barclay products. The Ramsbottom safety valves are mounted on the dome, while a tall stove-pipe chimney is fitted. A feature of Gibb and Hogg products is the provision of splashers over the driving wheels, while the open back to the cab will be noted, together with the converted wagon used as a tender. Photographed at Cardowan Colliery, Lanarkshire, in 1966.

33 Grant Ritchie, 272, built 1894

Grant Ritchie and Company were originally manufacturers of colliery winding equipment, but also built steam locomotives up to the 1920s. *No 61*, photographed at Fordell Colliery, Fife, in 1965, is of interest as it portrays a design favoured by all of the Kilmarnock manufacturers in the late nineteenth century. The stove-pipe chimney and safety valves on the dome provide a plain if somewhat austere appearance, while the port-hole window in the cab side was another recognisable feature.

34 Grant Ritchie, 805, built 1918

No 10 is a later product from Grant Ritchie and shows how in their development the products of this manufacturer copied the designs of Andrew Barclay. Typical Grant Ritchie features are the double-lever safety-valves on the dome and the single slide-bars of the motion. The cab has been fitted with a rather crude back-sheet, while this locomotive is also coupled to a tender though of more respectable design. *No 10* was photographed at Cardowan Colliery, Lanarkshire, in 1966.

35 Grant Ritchie, 536, built 1914

Several Scottish colliery companies favoured the 0-4-2 wheel arrangement, as this combined the flexibility of a four-coupled chassis with the power of a larger six-wheeled design. Lady Victoria Colliery, Midlothian, operated examples from both Andrew Barclay and Grant Ritchie. *No 7* from the latter manufacturer was photographed there in 1967. Of unspectacular design, the main Grant Ritchie features mentioned previously are present, while the built-up pattern of chimney is shown to advantage.

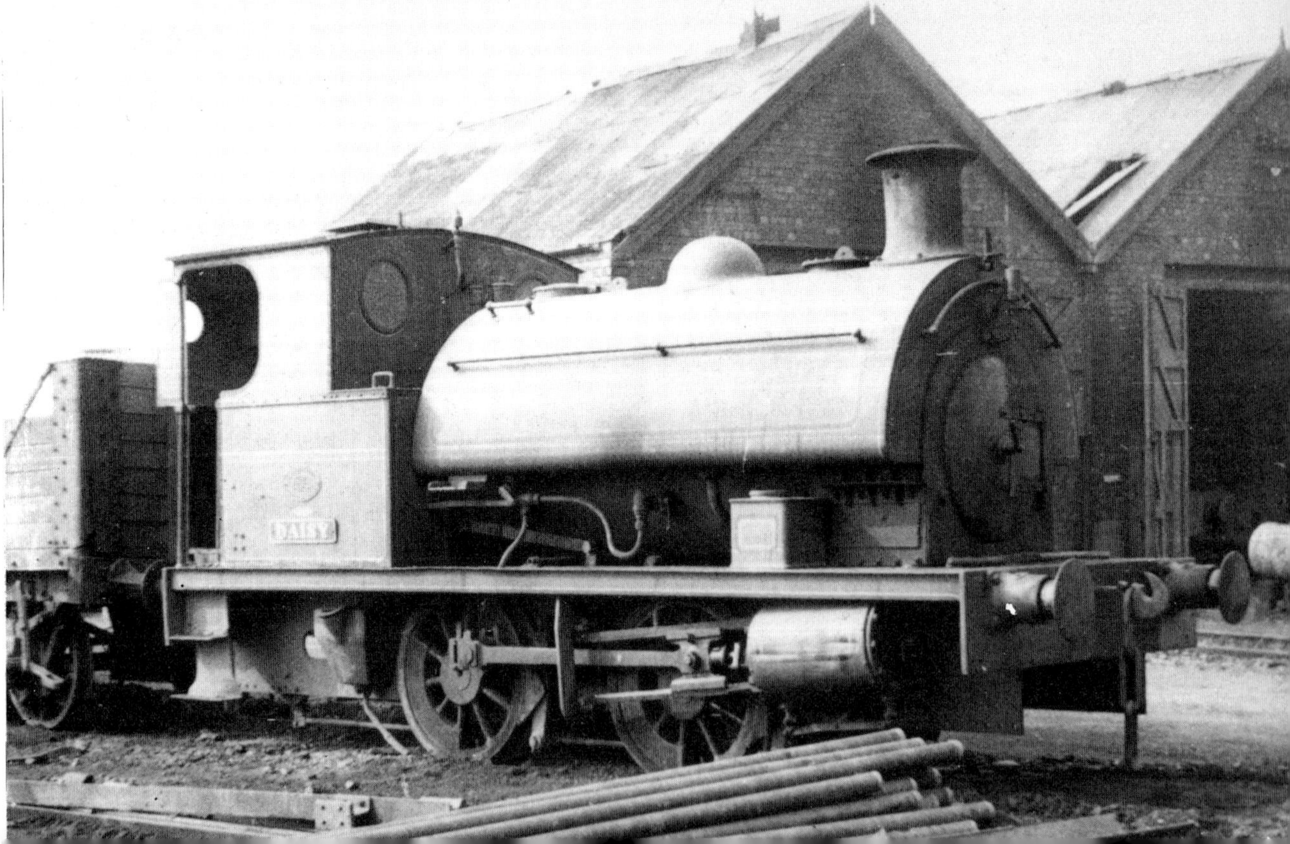
DAISY

36 (opposite, top) Hawthorn Leslie, 2800, built 1909

Met portrays the smaller standard gauge 0-4-OST produced by Hawthorn Leslie. Recognisable features are the round-section saddle tank with small dome and the Ramsbottom-pattern safety valves over the firebox. The fully enclosed cab and tall tapering chimney produce a well-balanced design of locomotive. Delivered new to Acton Lane Generating Station, Middlesex, *Met* was later transferred to Bow Generating Station, Essex, and was photographed there in 1966 devoid of name plates.

37 (opposite, bottom) Hawthorn Leslie, 2657, built 1906

Although built three years prior to *Met, Daisy's* modern appearance belies its age. It epitomises the most common Hawthorn Leslie four-coupled design in which the saddle tank covers the smokebox and boiler, but leaves the firebox exposed. The Ramsbottom safety valves are once more mounted over the firebox, while a wide-diameter chimney and small dome complete the picture. Delivered new to Blackwell Colliery Co Ltd, Derbyshire, *Daisy* was a spare engine at Bentinck Colliery, Kirkby in Ashfield, in 1964, but still in good external condition.

38 (below) Hawthorn Leslie, 3780, built 1932

Hawthorn Leslie and their successors developed a range of powerful four-coupled locomotives which combined a high tractive effort with the flexibility of a short wheelbase. South Durham Iron Company's Irchester pits latterly operated two example with 16in diameter cylinders, of which *Holwell No 30* was still operational in 1965. It is of standard Hawthorn Leslie design with saddle tank stopping short of the firebox and a large cab incorporating front coal bunkers. Note the smokebox wingplates, a feature favoured by this manufacturer during the 1930s. Photographed at Irchester ironstone pits, Northamptonshire, in 1965.

39 (above) Hawthorn Leslie, 3825, built 1934

When Corby Steelworks was enlarged in the 1930s further locomotives were required for shunting operations within the works. Hawthorn Leslie were called upon to supply thirteen locomotives which were to their standard six-coupled design with 16in x 24in cylinders and 3ft 8in diameter wheels. At first painted in standard green, they were later given a livery of yellow to make them more easily seen in the murky steelworks. Several were converted to burn oil, but later reverted to solid fuel when the price of oil increased. *No 12* was delivered new to Stewarts and Lloyds, Corby, in 1934 and is seen there in 1964.

40 (opposite top) Hawthorn Leslie, 3899, built 1936

Upton Colliery, Yorkshire, was the home of Hawthorn Leslie 3899, a larger machine than the Corby locomotive. Again a round-section saddle tank is fitted which covers the smokebox and is extended back to the cab front. The safety valves which are mounted over the firebox are therefore let into the saddle tank. Delivered new to Houghton Main Colliery Co Ltd, Barnsley, *Houghton No 4* was photographed in 1966 at Upton Colliery.

41 (opposite, bottom) Hawthorn Leslie, 3026, built 1913

In 1966 six-coupled side tanks built by three manufacturers could be found at Wellesley Colliery, Fife, including this example of one of Hawthorn Leslie's smaller products. It is an attractive design with unusual square-shaped front cab spectacle-plates. The boiler supports a tall dome and slim three-piece chimney while Ramsbottom-pattern safety valves are mounted over the firebox. Photographed outside the engine shed at Wellesley after the morning shift.

NATIONAL COAL BOARD
Nº 15
FIFE AREA

C B E S
No 5

No 27

42 (opposite, top) Hawthorn Leslie, 3904, built 1936

The West Midlands area of the Central Electricity Generating Board showed a preference for six-coupled side tanks for use in the Stafforshire and Warwickshire power stations. Hams Hall Station, Birmingham, employed no fewer than eight such locomotives which were delivered between 1936 and 1955. Of Hawthorn Leslie manufacture it has 18in diameter cylinders and shows this manufacturer's development of the locomotive illustrated previously. Delivered new to City of Birmingham Electric Supply Department, *CBES No 5* was photographed at Hams Hall in 1967.

43 (opposite, bottom) Hudswell Clarke, 1338, built 1918

This photograph of *NCB No 27* shows the main features of Hudswell Clarke's early 0-4-OST design. The saddle tank is more rounded than on locomotives built by the other Leeds manufacturers, but again leaves the smokebox exposed. A neat round dome is mounted centrally on the tank, while the safety valves are enclosed in the cab over the firebox. The smokebox sports a tall three-piece chimney and is flanked by the typical Hudswell Clarke design of sandbox. Delivered new to Earl Fitzwilliam's Collieries, Yorkshire, *No 27* was photographed at Elsecar Colliery, Yorkshire in 1966.

44 (below) Hudswell Clarke, 1696, built 1939

Even in the early 1960s the Oxfordshire Ironstone Company operated over twenty steam locomotives, of which twelve were six-coupled tanks and the remainder four-wheelers. Five of the latter were Hudswell Clarke products similar to *Jean* illustrated and were used to haul iron ore from the quarry to the tipping dock. Of standard Hudswell Clarke design the saddle tank is now full length and has the safety valves and whistle located on the dome. Originally supplied to the Royal Arsenal, Woolwich, and named *Beatty* it was obtained by the Oxfordshire Ironstone Company in 1958 and photographed at Banbury six years later.

45 (above) Hudswell Clarke, 1891, built 1961

The final steam-design produced by this manufacturer was an order of five locomotives for the National Coal Board West Yorkshire Area in the 1960s. The large cab has rounded eaves to the roof, while the safety valves are again mounted on the dome. The chimney is functional, though hardly attractive, and generally the locomotive has the appearance of a shortened Austerity tank. Delivered new to the NCB *H.C. No 3* is shown at Skiers Spring Colliery, Barnsley, where it was the working locomotive.

46 (opposite, top) Hudswell Clarke, 822, built 1912

Not surprisingly Hudswell Clarke supplied large numbers of locomotives to the various colliery companies in Yorkshire. Number 822 well portrays their standard saddle-tank design being little altered since built. The tall tapered chimney is a typical Hudswell Clarke feature, while the saddle tank appears taller than ever. Note the more respectable cab which is fitted. Photographed outside the running shed at Manvers Main Colliery, Yorkshire, where it was supplied new fifty-two years previously.

47 (opposite, bottom) Hudswell Clarke, 1308, built 1918

By the time *Rhos* was built, Hudswell Clarke had modified their designs to give a more modern appearance. The saddle tank has become rounder and now sports a dome. The chimney is shorter, while the smokebox door has an operating wheel rather than the customary handle. The crosshead-operated mechanical lubricator will also be noted. *Rhos* was supplied new to Lloyds Ironstone Company (later Stewarts and Lloyds), Corby, where it was photographed at Pen Green shed in 1965.

40

RHOS

B13041

50 (above) Hudswell Clarke, 1877, built 1954

Hudswell Clarke supplied four locomotives of this design to the National Coal Board in Nottinghamshire. The name is derived from East Midlands Division Area 4 to which the locomotive was allocated. The modern-style cab has a squat appearance, which is accentuated by the high side-tanks. Nevertheless it retains all the features characteristic of Hudswell Clarke designs. Pictured at Brookhill Colliery, Kirkby in Ashfield, Nottinghamshire, in 1965 where it was later scrapped.

48 (opposite, top) Hudswell Clarke, 607, built 1903

A visit to Corby steelworks in 1964 was rewarded by this photograph of *Pen Green.* Even though it was more than sixty years old it had escaped drastic rebuilding, although the original chimney had been replaced by one of more austere pattern and Ross pop safety-valves had been fitted. The large sandboxes mounted on the running plate either side of the smokebox are another recognisable Hudswell Clarke feature. *Pen Green* was delivered new to Lloyds Ironstone Company, Corby, where it was later broken up.

49 (opposite, bottom) Hudswell Clarke, 1419, built 1923

Part of the Oxfordshire Ironstone Company's system was a double-track 'main line' from the loading docks at Wroxton to the Great Western exchange sidings north of Banbury. The severe gradients necessitated the use of six-coupled locomotives and Hudswell Clarke supplied three side-tanks to the design illustrated. *The President* incorporates all the trademarks of this manufacturer and interesting features are the sharply-angled outside cylinders and the high side-tanks with inspection holes. Photograph taken at Banbury in 1964.

Nº 9
Nº 9
FIFE AREA

62

51 (opposite, top) Hudswell Clarke, 859, built 1909

A less-common design of side tank from Hudswell Clarke is portrayed by *No 9* of the NCB Fife area. The tall side-tanks extend almost to the smokebox front and have been cut away to allow access to the inside motion. This has necessitated a separate splasher for the leading wheel, while the sand boxes have been positioned below the footplate, the rear one presumably being refilled from inside the cab. Note also the curved handrail on the smokebox front and the cover to the inside cylinders. Photographed in 1966 at Wellesley Colliery, Fife.

52 (opposite, bottom) Hudswell Clarke, 940, built 1911

The Manchester Ship Canal Company standardised almost exclusively on inside cylinder, six-coupled side tanks, of which the majority were supplied by Hudswell Clarke. Typical of the later design of locomotives is *No 52.* The side tanks extend far forward and slope downwards to improve the forward view. Ross pop safety-valves are mounted over the firebox, while a tall dome and tapering chimney complete this pleasing design. Delivered to MSC in 1911, *No 52* was photographed at Mode Wheel Shed in 1965.

53 (below) Hunslet Engine Company, 317, built 1883, 2ft. 0in. gauge

Hunslet built a large number of narrow-gauge locomotives for industrial use and typical of those delivered to the North Wales Slate quarries is *Lilian.* It has the standard Hunslet-pattern saddle tank with tall chimney and round dome, while the safety valves are enclosed in a brass casing. This example is cabless but others were fited with fully enclosed cabs. Also to be noted is the dropped footplate at the rear end and the outside frames, necessary because of the narrow gauge. *Lilian* was delivered new to Lord Penrhyn's Slate Quarry, Caernarvonshire, where it was photographed in 1965.

55 (opposite, top) Hunslet Engine Company, 1493, built 1925

A later design of 0-4-OST is illustrated by *NCB No 11.* The angular saddle-tank has given way to one of round cross-section, which again stops short of the smokebox. This example sports a short chimney and is cabless because of loading-gauge restrictions at the colliery. New to James Oakes and Co, Ltd, for use in their Pye Hill Colliery, it is now preserved at Shackerstone, Leicestershire, where it was photographed in 1974.

56 (opposite, bottom) Hunslet Engine Company, 1557, built 1927

Jessie illustrates a four-coupled side tank from this Leeds manufacturer. It has 16in. diameter cylinders and is fitted with Walchaerts valve gear, an unusual feature for industrial locomotives. The sandboxes mounted on the smokebox will be noted, as will the coal bunker fitted to the cab rear. Built for the Tyldesley Coal Co, Lancashire. *Jessie* was photographed at NCB Central Workshops, Walkden, in 1966.

54 (below) Hunslet Engine Company, 1403, built 1920

Despite its ancient appearance this locomotive was built as late as 1920. The design however is similar to products from Hunslet Engine Company some thirty years earlier. The flat-sided saddle tank covers only the boiler barrel and has a neat, round, dome placed centrally. A modern cab has replaced the curved weather board found on early examples, while spring balance safety-valves have given way to ones of Ross pop design. Delivered new to John Knowles Ltd, it was photographed at that firm's Woodville Works, Derbyshire, in 1966.

NCB
TASKERS
TRAILERS

TREASURER

59 (above) Hunslet Engine Company, 3872, built 1958

In 1923 the Hunslet Engine Company produced the first of their standard 16in. cylinder class which was to be built in numbers over the following thirty-five years. Recognisable features are the round-section saddle tank stopping short of the smokebox, but extending to the cab front. The safety valves and whistle are mounted on the dome, while the chimney is a typical Hunslet one-piece casting. The commodious cab now has a rear coal bunker with a sharply sloping back, compared with the forward side bunkers shown in Fig. 58. *Frank* was the last of the class to be built and was supplied to the Oxfordshire Ironstone Company, Banbury, where it was photographed in 1964.

57 (opposite, top) Hunslet Engine Company, 1446, built 1929

Treasurer illustrates a six-coupled design with outside cylinders. It is based on Hunslet's 16in. design, but the overall wheelbase has been reduced and the 15in. diameter cylinders placed outside the frames together with Walchaerts valve gear. The whistle and safety valves are again located on the dome, while the usual Hunslet cab and bunker is fitted. Delivered new to Oxfordshire Ironstone Co Ltd, Banbury, *Treasurer* was transferred to Stewarts & Lloyds, Corby, in 1937 and was photographed outside Corby Steelworks shed in 1965.

58 (opposite, bottom) Hunslet Engine Company, 498, built 1890

A good example of the standard 0-6-OST produced by Hunslet towards the end of the nineteenth century for contract and industrial use. The angular flat-sided saddle tank with round dome, favoured by this manufacturer is shown to advantage. Note the half-round smokebox door hinged at the top, and the wheel design, both indications of an elderly locomotive. This example was delivered new to Griff Colliery Company Ltd, Nuneaton, and was transferred in NCB days to Haunchwood Colliery where it was photographed in 1966 stripped of nameplates.

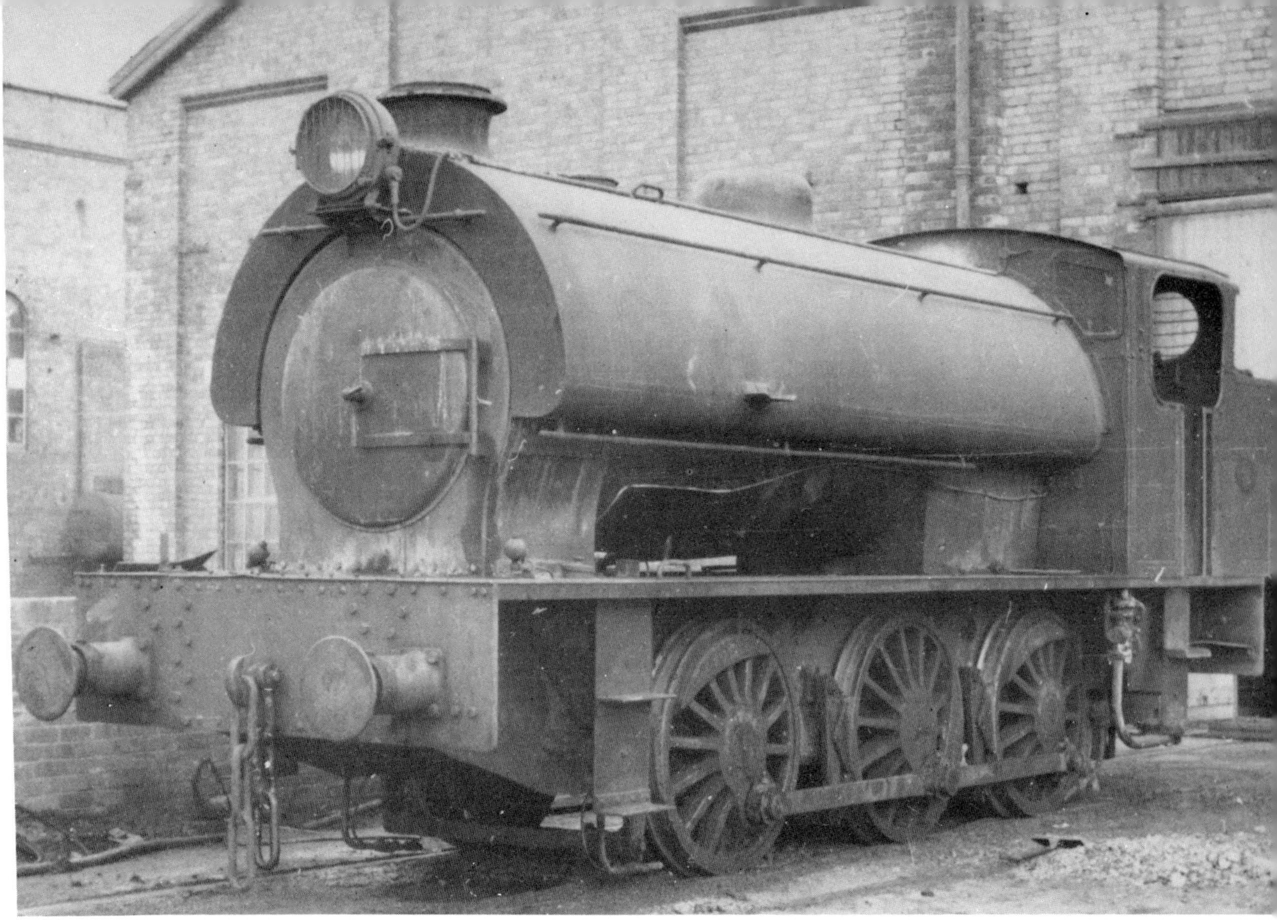

60 Hunslet Engine Company, 2882, built 1943

The industrial design produced in greatest numbers is without doubt the Austerity saddle tank. Its origin can be found in Hunslet's 18in. design which with slight modification was chosen as the standard shunting locomotive for the Ministry of Supply in World War II. The round full-length saddle tank is topped by a flared chimney and a flat-topped dome, while the safety valves are mounted over the firebox. The back of the coal bunker is vertical and overhangs the rear buffer beam which is necessarily fitted with extended buffer shanks. The example illustrated was supplied new to the WD as their 75035 and was sold to the NCB East Midland Division in 1947. Photographed at Coppice Colliery, Ilkeston, in 1964.

61 (opposite) Hunslet Engine Company, 1475, built 1924

In 1924 the Hunslet Engine Company supplied two locomotives to the Bridgewater Collieries, Lancashire. They were large side-tanks with six-coupled wheels and 18in. diameter outside cylinders. Two unusual features for an industrial design are the Walschaerts valve gear and Belpaire firebox. Note also the safety valves and whistle mounted on the dome. They both passed into NCB ownership on nationalisation and *Bridgewater*, the second of the pair, is seen in 1962 at Hill Top Colliery, six years before being broken up at Walkden area workshops.

63 (opposite, top) Kerr Stuart, 3123, built 1918

A standard gauge Kerr Stuart four-coupled saddle tank is portrayed by *Kearsley.* Delivered new to the Inland Waterways & Docks Board it was acquired by the Clifton and Kersley Coal Company and passed into NCB ownership on nationalisation. It is photographed at Walkden area workshops, Lancashire, in 1966 out of use. It will be noted that the round saddle-tank extends full length and the safety valves are located on the centrally-mounted dome.

64 (opposite, bottom) Kerr Stuart, 4034, built 1920, 2ft. 6in. gauge

Superior represents a six-coupled design from this manufacturer. The frames are once again located outside the driving wheels, which are close coupled to afford a short wheelbase. The side tanks extend to the smokebox front and are cut away to allow access to the inside motion and perhaps to reduce weight. A tall parallel chimney is fitted, while the sandboxes are mounted on the boiler which is domeless. New to Edward Lloyd Ltd, Sittingbourne, Kent where it was photographed in 1964.

62 (below) Kerr Stuart, 1049, built 1908, 2ft. 6in. gauge

A popular Kerr Stuart design was the Brazil class which had the uncommon 0-4-2 wheel arrangement. The round saddle-tank covers only the boiler barrel and is devoid of dome. On this example the flat-sided smokebox has a spark-arresting chimney, but a conventional design was more commonly fitted. Similarly, while on this locomotive the wheels are placed inside the frames, they could be placed outside according to the track gauge. *Premier* was the first of four such locomotives delivered to Edward Lloyd's paper mills, Sittingbourne, Kent and was photographed there in 1964.

SUPERIOR

65 Kerr Stuart, 3068, built 1917

In 1917 Kerr Stuart supplied ten six-coupled side tanks to the Inland Waterways and Docks Board. They formed the 'Victory' class and had 17in. diameter cylinders and 4ft. 0in. driving wheels. They were soon disposed of after cessation of hostilities and many were bought by colliery companies. *Francis* was purchased by the Earl of Bridgewater Manchester Colliery Company for use at Astley Green Colliery, Lancashire, where it was photographed in 1962. Note the Belpaire firebox, an unusual feature on industrial locomotives.

66 Kitson, 5158, built 1915

Kitsons of Leeds supplied two locomotives to Walsall Wood Colliery, Staffordshire, to the design illustrated here. It is a typical Kitson six-coupled saddle tank in which the tank covers only the boiler barrel. A tall tapering chimney is fitted, while the safety valves are mounted over the firebox. *Lord Kitchener* spend its whole life at the colliery and was photographed at work there in 1964.

67 Kitson, 5358, built 1921

Also seen in the Cannock Chase area is a Kitson side tank, examples of which were supplied to the Manchester Ship Canal Company and the Lambton, Hetton and Joicey Colliery Company. The sharply-sloping side tanks are a feature of this design as are the tall, tapered chimney, the cover to the inside cylinders and the way in which the front of the side tanks merges into the smokebox. No 5358 was new to the Glyncorrwg Colliery Company, Glamorgan, and later sold to the Coppice Colliery Company, Staffordshire. It is seen out of use at the NCB Area Workshops at Chase Terrace, Staffordshire, in 1964.

68 Kitson, 2509, built 1883

Many manufacturers and operators in early days favoured the long-boiler design of locomotive in which the firebox was located behind the rear axle, rather than between the centre and rear wheels. This allowed a short wheelbase and produced a much more adaptable locomotive. The Consett Iron Company operated nineteen locomotives supplied new to this design between 1883 and 1941 for use in its steelworks and collieries. Twelve were transferred to NCB ownership and *No 41* was the last survivor, having been rebuilt with usable parts from its sister locomotive *No 42.* Photographed at Derwenthaugh, County Durham in 1969 *No 41* survived to be acquired for preservation.

N. C. B.

69 (opposite) Kitson, 4263, built 1904

A locomotive from the Lambton, Hetton and Joicey Colliery Company's stud is seen in the form of *No 29*. It was built by Kitson of Leeds in 1904 and is an attractive and well-proportioned locomotive. An interesting feature is the combined leading splasher and sandbox which merges into the smokebox wingplates. Note also the sharply rounded cab-roof, a feature of many Lambton locomotives because of restricted loading gauge. *No 29* is seen in 1969 at Philadelphia, Durham.

70 Stephen Lewin, built 1863

A manufacturer based at Poole in Dorset was Stephen Lewin, who in the 1860s and 1870s produced several well-tank designs for industrial use. *No 18* was originally a well-tank but in 1923 acquired the saddle tank and chimney from a withdrawn Black Hawthorn engine. The sharply-inclined cylinders and round-section piston rods are features distinguishing Lewin's products. Although out of use for several years, *No 18* was later returned to traffic, and when photographed in 1966 at Seaham Harbour, County Durham, it was by far the oldest locomotive in industrial use in this country.

71 Manning Wardle, 1877, built 1915, 2ft. 6in. gauge

Typical of one of Manning Wardle's narrow gauge designs is *Chevallier* which depicts a class of which most representatives were delivered overseas. It has 12in. x 15in. cylinders and is fitted with Walchaerts valve gear. The spring balance safety-valves are unusual for a Manning Wardle locomotive of this period, while the boiler-mounted sandboxes will also be noted. Delivered new to the Admiralty for use on the Chattenden and Upnor Railway, *Chevallier* was sold to Bowater Lloyd's paper mills in 1950 and was photographed at Sittingbourne in 1964.

72 (opposite) Manning Wardle, 865, built 1882

The classic outline of the early Manning Wardle product is shown to advantage by *Aldwyth*. The vertically-hinged half section smokebox door is of interest, together with the design of driving wheels found on older locomotives from this manufacturer. Although the brass safety-valve casing has been removed, the spring-balance valves are retained and are mounted ahead of the sparse cab. Believed to have been supplied new for contract work, *Aldwyth* eventually passed to the Air Ministry (later Ministry of Public Works and Buildings), Kidbrooke, where it worked until the early 1970s. Photographed at Kidbrooke in 1966.

73 Manning Wardle, 1795, built 1912
No 14 illustrates the Manning Wardle standard gauge 0-4-OST with flat-sided saddle tank, stopping short of the smokebox, topped by a tall three-section chimney. The safety valves on this example are mounted over the firebox and enclosed in the cab, but earlier products had a polished brass casing located ahead of a curved weather board. *No 14* is seen at the Irchester pits of the South Durham Iron and Chemical Company, Northamptonshire, in 1964 having been acquired from the Wensley Lime Company, Yorkshire, in 1957.

74 Manning Wardle, 1601, built 1903
Manning Wardle locomotives were popular with the cement works and quarries of South-East England and several were to be found, still at work in the 1960s. The last survivor was *Arthur,* which was a standard class L locomotive having 12in. x 18in. cylinders. The design has been tidied up and generally modernised, but nevertheless the similarity to *Aldwyth* can be seen. *Arthur* was obtained from P. & W. Anderson and Company, contractors, in 1923 and was photographed at APCM Kent Works in 1966.

75 Manning Wardle, 1762, built 1910

Corby Steel Works had the distinction of operating locomotives to this design supplied by three successive manufacturers. Manning Wardle supplied this example in 1910, while Kitson supplied a further seven in the 1930s. In 1940 Robert Stephenson and Hawthorns, who had in turn acquired the patterns, produced a further five locomotives with little change in the original design. They were impressive locomotives with 16in. x 22in. cylinders and 3ft. 6in. diameter wheels. *Dolobran* is seen at Pen Green shed, Corby, in 1965.

GEORGE SLATER LTD
Nº 3

B.B.S.LTD
Nº4
Nº
4

76 (opposite, top) Markham, built 1912

Markham and Company were manufacturers of colliery winding machinery based at Chesterfield, Derbyshire. In 1889 they commenced steam locomotive production and up to 1914 they built nineteen locomotives. A characteristic of their products was the angular saddle tank with its multitude of round-headed rivets. Also favoured was the tapered chimney and safety valves set on a centrally-mounted dome. George Slater's example was originally supplied to the English Steel Company's River Don Works, Sheffield and sold to Slater in 1932. It was photographed out of use in 1967 shortly before being broken up.

77 (opposite, bottom) Nasmyth Wilson, 454, built 1894

Brown Bayley and Company owned one of the smaller steelworks in Sheffield, which was located adjacent to the ex-Great Central Railway line from Sheffield to Manchester. They were great supporters of steam traction, using Sentinel steam lorries for internal road movement and several steam locomotives for rail traffic. The oldest locomotive operated was also in later years the last surviving product from Nasmyth Wilson. Of standard design, the dome would originally have been fitted with spring balance safety-valves, but these were later replaced. The unusual shape of the cab side-sheets is another feature which is typical of Nasmyth's industrial saddle-tank designs. Supplied new to Brown Bayley and photographed at their works in 1968.

78 (below) Neilson, 3629, built 1887

Although Neilsons of Glasgow made their reputation in the main-line locomotive field, they nevertheless supplied a considerable quantity of machines for industrial use. One of their standard designs is illustrated here and incorporates several interesting features. Typical of the manufacturer is the tapered stove-pipe chimney and the unusually-shaped cab front with spectacle plates set necessarily close together. The curvaceous saddle tank with safety valves on the dome and the single slide-bars will also be noted. *No 3* is seen at Bairds and Scottish Steel Company's Gartsherrie Steelworks, Coatbridge, in 1965.

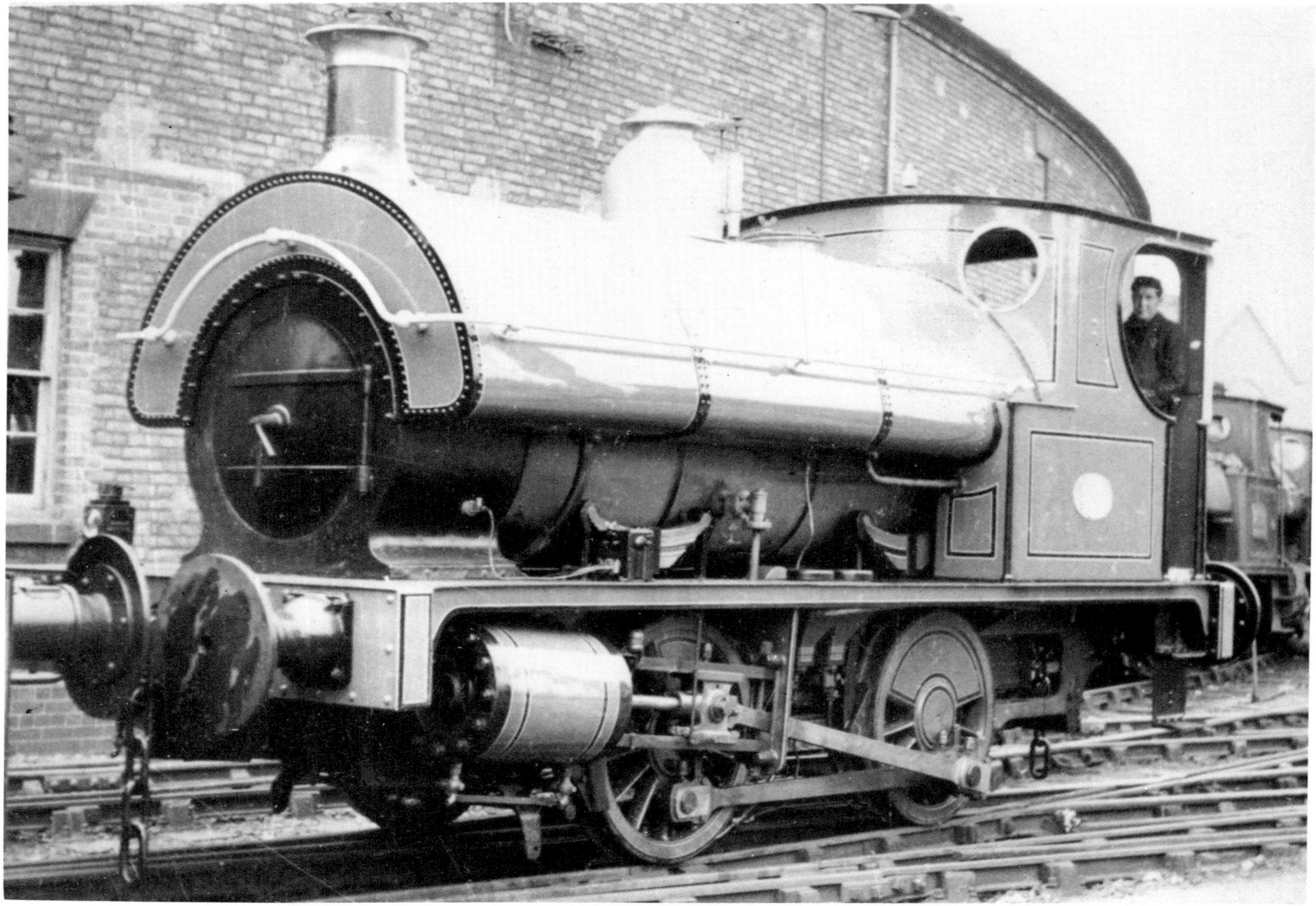

79 Neilson Reid, 5760, built 1900

The town of Burton on Trent is best known nationally for its beer and to the local inhabitants for the numerous level crossings associated with the brewery railways. The largest system was that of Bass Ratcliffe and Gretton Ltd, who operated no fewer than twelve locomotives similar to the design illustrated. *No 2,* which was built by Neilson Reid in 1900, shows how well proportioned and pleasing to the eye the locomotives were. With their turkey-red livery and polished brasswork they were a fine sight. Photographed at Burton on Trent in 1963.

80 North British, 18385, built 1908

In 1966 Bairds and Scottish Steel Company Ltd still operated a full steam fleet of six locomotives at their Gartsherrie Works, of which the newest was built in 1908. *No 19* of their stud depicts the North British Locomotive Company development of the Neilson design shown previously. The main features are still present, but the locomotive now sports a more respectable fully-enclosed cab. Note the dumb buffers fitted to avoid buffer lock on the sharp curves. New to Bairds and Scottish Steel Company, *No 19* is seen at work in Gartsherrie Works in 1965.

81 North British, 16463, built 1904

An elderly North British Locomotive Company product was still at work at Wellesley Colliery, Fife, 1966. It has a round-section full-length saddle tank with safety valves mounted over the firebox, but let into the tank. *No 7* is fitted with a commodious cab with flared coal-bunker and once again sports a stove-pipe pattern chimney. Note that it has been fitted with a new front driving-wheel of different design from the rest.

82 (above) North British, 24564, built 1939

Coventry No 1 illustrates a side-tank design from the same manufacturer. Similar locomotives were built by the predecessors of NBL for use at Scottish Collieries and while it is an industrial design one would be forgiven for thinking it to be an ex-main line locomotive. It is interesting to note the height of the side tanks, which have inspection holes cut in them to afford access to the inside motion. Ramsbottom-pattern safety valves are fitted, while a neat dome and chimney design complete a well-proportioned locomotive. It was delivered new to the Warwickshire Coal Company in 1939 and was transferred in NCB days to Haunchwood Colliery, where it was photographed in 1966.

83 (opposite, top) Peckett, 1258, built 1912

Thomas Peckett's locomotives were renowned for their reliability and robustness, and it is notable that his original designs of the 1880s were varied only in detail during the following seventy years of production. *Rothwell* exemplifies the early Peckett R2 class. Features are the built-up chimney, tall brass bell-shaped dome and smokebox wingplates. One of a pair with 12in. x 18in. cylinders delivered new to James Pain Ltd, *Rothwell* was at first allocated to Glendon North Ironstone Quarries, but later transferred to Irchester pits, Northamptonshire, where it was out of use in 1964.

84 (opposite, bottom) Peckett, 1823, built 1933

The Peckett M5 class was perhaps the most aesthetically pleasing of that manufacturer's standard-gauge designs, with a tall tapering chimney and other features in proportion. It had small 2ft. 6in. diameter wheels while 10in. x 15in. cylinders were fitted. The example illustrated was delivered new to the Birmingham, Tame and Rea District Drainage Board and sold to Ambrose Shardlow and Company by George Cohen Ltd. It was photographed among the greenhouses at Shardlow's Meadowhall Works, Sheffield, in 1964.

ROTHWELL

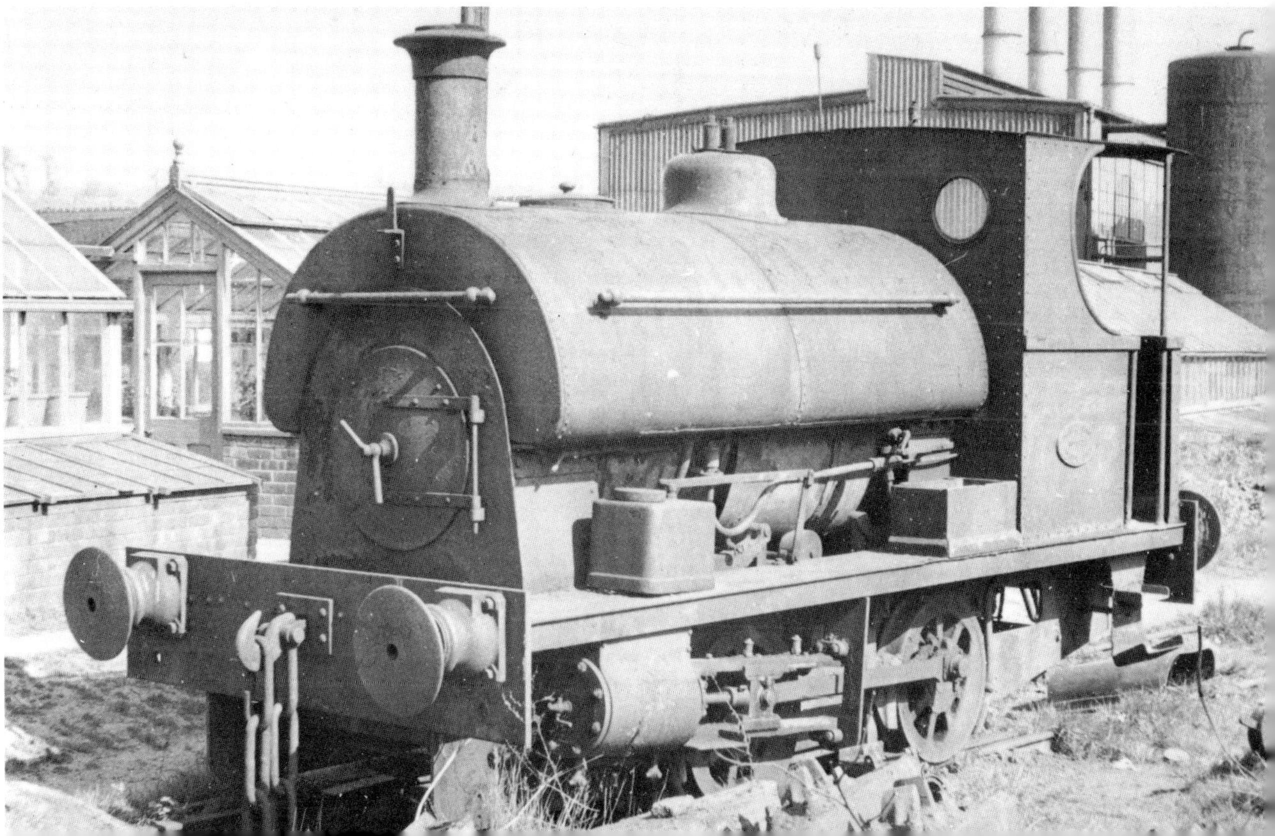

85 (above) Peckett, 1880, built 1935

One of the most numerous classes built by Peckett and Sons was R4, of which fifty-six representatives were constructed. The class had 12in. x 20in. cylinders and 3ft. 0in. diameter wheels and the example illustrated here shows the features favoured by Peckett in later years. The curved smokebox wingplates have given way to ones of straight design, while Ross pop safety-valves are enclosed in a smaller dome cover. *Nelson* is seen in the works at Greenhithe, Kent, in 1966 having been supplied new to Albert E. Reed's Aylesford paper mills, New Hythe, and later transferred.

86 (opposite, top) Peckett, 2119, built 1950

Stewarts and Lloyds used this W7 class locomotive at their Bromford Bridge Tube Works, Birmingham. Although a much larger machine, having 14in. x 22in. cylinders one can see the family resemblance to the Peckett illustrated previously. Peckett introduced their W7 class in the early 1940s and built a total of fifty-two locomotives, many of which were supplied to the Royal Ordnance factories. Number 2119 however was delivered new to Stewarts and Lloyds, Bromford Bridge, and photographed there in 1966 twelve months before being broken up.

87 (opposite, bottom) Peckett, 2029, built 1942, metre gauge

Passengers travelling between St Pancras and the East Midlands may have noticed just north of Wellingborough Station a narrow-gauge line passing underneath the main line and at right angles to it. Sometimes they were rewarded with a glimpse of a saddle tank taking its rake of wagons up the gradient to the east side, towards the quarry. This was the Wellingborough Iron Company's line later acquired by Stewarts and Lloyds. They had three locomotives built by Peckett and Sons, of which *No 87* was the last to be supplied in 1942. It was nominally based on Peckett's standard-gauge class R4 0-4-OST but modified to take a metre gauge, six-coupled chassis. Note the centre buffer-coupling and how the cab is narrower than the saddle tank. Photographed in the exchange sidings at Wellingborough in 1964.

88 Peckett, 1378, built 1914

The early six-coupled saddle-tank design shown here retains the original dome and curved smokebox wingplates, while the fully-enclosed cab now sports a rear coal-bunker. In contrast to contemporary manufacturers, Pecketts fitted a respectable cab as a standard feature on their early products. *Westminster* represents class B2 having 14in. x 20in. cylinders and was delivered new to Alpha Cement Company, Shipton on Cherwell, Oxfordshire, where it was photographed in 1967.

89 Peckett, 1787, built 1933

Newdigate Colliery, Warwickshire, was the home of two representatives of Peckett's class X2, of which a total of fifty-two locomotives were built. They had 16in. x 24in. cylinders with 3ft. 10in. diameter wheels and were powerful and impressive locomotives. The characteristic Peckett features are present, while it will be noted that the cab is fitted with a flared coal-bunker. *No 4* was delivered new to Newdigate Colliery, and is pictured there in NCB days in 1967.

90 Peckett, 1745, built 1927

The Warwickshire Coal Company showed a preference for Peckett locomtives, taking delivery of three similar machines between 1924 and 1927. One of them is illustrated here and is of Peckett class OX1. Built with 16in. x 24in. cylinders it may be compared with *Westminster* to appreciate the development of Peckett's designs. The later style of straight smokebox wingplates and dome cover will be noted. Delivered new to the Coventry Coal Company in 1927, *Coventry No 4* was photographed at Arley Colliery, Nuneaton, in 1967 resplendant in NCB blue livery.

91 (above) Sentinel, 6515, built 1926

In the 1920s Sentinel of Shrewsbury started production of chain-driven locomotives having a vertical boiler in the cab. They were successful within limitations, for in practice the chain was prone to breaking if very heavy loads were pulled. Nevertheless many industrial operators used these locomotives, while the main-line railway companies all took delivery of Sentinel products, the LNER collection totalling fifty-five examples. *Isebrook* depicts the early design which is somewhat boxlike in appearance. It was originally supplied to the Great Western Railway, but returned to the makers who resold it to Thos E. Gray, Burton Latimer, Northamptonshire, where it was photographed in 1966.

92 (opposite, top) Sentinel, 9369, built 1946

Thomas Gray took over the Isham ironstone quarries of Lloyds Ironstone Company in 1925 and quarried silica sand. The first locomotives were of conventional design but one was later rebuilt with a Sentinel lorry boiler and thus instituted what was to be an exclusive use of Sentinel locomotives. *Musketeer* is of Sentinel's post-war streamlined design and is one of a pair owned by Grays. The curvaceous casing makes an extremely attractive locomotive, which, in the author's opinion, is more pleasing than the standard design in the next illustration. Delivered new to Williams and Williams Hooton, *Musketeer* was later acquired by Thomas E. Gray and photographed at Burton Latimer, Northamptonshire in 1966.

93 (opposite, bottom) Sentinel, 9552, built 1953

NCB No 50 represents the standard postwar 200hp design produced by Sentinel. It is recognisable by the angular superstructure and modern design of cab, which on this example has a stepped footplate. The chassis has the customary outside frames with roller-bearing axleboxes, and the overall appearance is not unlike some of the diesel locomotives produced at the time. Supplied new to the NCB, West Yorkshire Division, *No 50* is seen out of use at Manvers Main Colliery, Yorkshire in 1964.

MUSKETEER

No 42

94 (opposite, top) Robert Stephenson, 3801, built 1920

The Lambton, Hetton and Joicey Colliery Company was one of the largest and best-known colliery systems in North-East England. It connected the various collieries with the staithes at Sunderland, part of which journey was made over the North Eastern Railway Company's metals. It was necessary to have locomotives of main line standards for this purpose and a large fleet of six-coupled side tanks was built up. Typical of the locally-produced locomotives is *No 42* seen hauling a rake of 21 ton hopper-wagons towards Lambton plant. It was supplied new by Robert Stephenson to the Lambton Hetton Coal Company in 1920 and was broken up about 1969.

95 (opposite, bottom) Robert Stephenson & Hawthorns, 7818, built 1954

Castle Donington power station has employed steam locomotives for internal shunting since it was commissioned. Two identical locomotives were supplied by Robert Stephenson and Hawthorns, one of which is seen here in 1965. Having 16in. diameter cylinders it shows the progressive development of the four-coupled saddle tank since the early twentieth century. Note the extended coal space ahead of the cab and the re-railing beam across the frame to prevent the locomotive falling completely between the rails in the event of a derailment.

96 (below) Robert Stephenson & Hawthorns, 6947, built 1938

Robert Stephenson's design of 0-6-OST for steelworks use is illustrated. It has a robust appearance with its short wheelbase chassis with outside cylinders. The saddle tank covers only the boiler barrel, while an attractive tapered chimney tops the large-diameter smokebox. Supplied new to Appleby Frodingham Steelworks, *39* was sold to the National Coal Board when displaced by diesel locomotives and was photographed in 1965 at Gedling Colliery, Nottinghamshire.

97 Robert Stephenson & Hawthorns, 7688, built 1951

In the early 1950s Robert Stephenson and Hawthorns supplied a large number of locomotives to the North-Eastern collieries as part of a fleet replacement programme, and *No 4* seen here at Philadelphia, County Durham, in 1966 exemplifies one of the outside-cylinder designs. The massive superstructure seems disproportionate to the chassis, with its small driving wheels and short wheelbase. The austere cab features with square corners to the side cut-out will be noted as will the square coal-bunker.

98 Robert Stephenson & Hawthorns, 7298, built 1946

While Hawthorn Leslie and Robert Stephenson and Hawthorns will be best remembered for their various outside-cylinder designs, they nevertheless did produce some with inside cylinders. One of their six-coupled products is illustrated here. *Progress* follows traditional Hawthorn Leslie designs, with a full length round-section saddle tank, and a fully-enclosed cab with flared coal-bunker. The safety valves are let into the tank over the firebox and a round dome and flared chimney are fitted. Delivered new to the Moira Colliery Company, Leicestershire, *Progress* was later transferred to Cadley Hill Colliery, Derbyshire, where it was photographed in 1970.

99 Robert Stephenson & Hawthorns, 7668, built 1950

For its final class of steam locomotives Stewarts and Lloyds again turned to Robert Stephenson and Hawthorns. Once more they produced a special design peculiar to Corby, but based on the Austerity saddle tank. The overall length and wheelbase has been increased and the wheels reduced to 4ft. 0in. diameter. The saddle tank stops short of the firebox, while the cab has the angular side cut-out typical of later RSH products. It is interesting that of all its locomotives at Corby, Stewarts and Lloyds never had a standard Austerity design. The nearest was the Hunslet 18in. locomotive from which the Austerity was developed. *No 57* is seen at Corby in 1966.

100 Robert Stephenson & Hawthorns, 7765, built 1954

No 40 was one of a pair of side tanks with outside cylinders supplied to the NCB at Ashington, Northumberland, and photographed there in 1966. It has a long wheelbase with somewhat large wheels, and sanding gear is applied to the front and rear wheels only. The cab and bunker are standard RSH practice of the day, while the side tanks have a sharply-sloping top to improve forward visibility.

101 Robert Stephenson & Hawthorns, 7609, built 1950

In 1950 Robert Stephenson and Hawthorns supplied three locomotives identical to *No 31* illustrated. This is a side-tank version of the Austerity saddle-tank design and has 18in. diameter cylinders. The tall side-tanks are cut away to facilitate inspection of the motion, while the cab has rectangular front spectacle-glasses. The smokebox incorporates an inside cylinder cover and is topped by an attractive flared chimney. Delivered new to NCB Ashington, where it was photographed in 1966.

102 Yorkshire Engine Company, 2345, built 1934

2345 is typical of the Yorkshire Engine Company's later four-coupled locomotives. Identification features are the round-section, full-length saddle tank with squat boiler-mountings and safety valves on the dome. The smokebox is provided with a separate saddle, while the cab has a sharply-sloping back. It was one of a number of similar locomotives supplied to Steel Peach and Tozer Ltd, Sheffield, but was later transferred to the English Steel Company's slag disposal point at Woodhouse Mill, where it was photographed in 1967.

103 Yorkshire Engine Company, 327, built 1882

Another veteran surviving at Corby Steelworks until 1964 was *No 25.* The unusual ogee-shaped saddle tank and solid wheels are a sure sign of age. The cab is a replacement for the original weather-board, while the Ross pop safety-valves and chimney are also later additions. It was delivered new to contract passing to Henry Heys Quarries, Lancashire, and was transferred to Stewarts & Lloyds, Corby, in 1945.

104 Yorkshire Engine Company, 2197, built 1928

By the 1920s the Yorkshire Engine Company had adopted a round-section saddle tank in preference to that shown in Fig. 103 and *Bentinck No 3* illustrates the six-wheel design produced at that time. The saddle tank extends full length with the safety valves mounted on a squat dome, while a slim chimney completes a pleasing locomotive. *Bentinck No 3* was delivered new to the New Hucknall Collieries Company Ltd and later transferred to Bentinck Colliery, Kirkby in Ashfield, where it was photographed in 1963.

105 Yorkshire Engine Company, 2413, built 1943

Performing shunting operations at Stanton Ironworks in 1964 was *Sheepbridge No 28*. Although a much larger machine than *Bentinck No 3*, the basic family likeness can nevertheless be seen. The boiler mountings have been reduced in height, while the cab has the typical sloping back to the coal bunker. The operating wheel instead of a handle on the smokebox door is a feature favoured by the Yorkshire Engine Company. *Sheepbridge No 28* was delivered to the Sheepbridge Coal and Iron Company, but was later transferred to Stanton Ironworks, Derbyshire.

Industrial Locomotives in Action

106 (below) Andrew Barclay, 705, built 1891

An elderly Andrew Barclay tank is seen at Pye Hill Colliery, Derbyshire, in 1967 against the brickworks. In common with all Pye Hill locomotives it has had the cab and boiler fittings reduced to meet the restricted loading gauge. Formerly in contract ownership, it was acquired by the Mapperley Coal Company and was transferred to Pye Hill in 1957 in NCB days. Note the rectangular builder's plate favoured by Andrew Barclay at the time.

107 (opposite, top) Peckett, 1891, built 1940

Seen shunting empty coal wagons at Manvers Main Colliery, Yorkshire, in 1965 is *Wilf.* It is a standard Peckett X2 locomotive, which was delivered new to Manvers Main Colliery in 1940. It is unusual in having the early pattern of smokebox wingplates, which may have been due to an exchange of boilers with a sister locomotive built in 1921.

108 (opposite, bottom) Robert Stephenson and Hawthorns, 7049, built 1942

The Cumberland Coalfield was an isolated strata found around Whitehaven. The various collieries were originally owned by the Earl of Lonsdale, later becoming the Cumberland Coal Company, a subsidiary of the Coltness Iron Company, and on nationalisation they formed the Cumberland Area of the NCB Northern Division. The picture was taken at William Pit, Whitehaven, in 1966 where the locomotive was shunting empty coal hopper-wagons. It is of standard Hawthorn Leslie design and was originally supplied to the Distington Engineering Works, Distington, Cumberland. They sold it in 1950 to the Workington Iron and Steel Company, from whom the NCB acquired it in 1963.

N.C.B.

109 Andrew Barclay, 1338, built 1913 and 1785, built 1923

A brace of Andrew Barclay side tanks is seen on the former Dalmellington Iron Company's system in Ayrshire. On the right is *No 17,* which was the first six-coupled locomotive obtained by Dalmellington and has 18in. x 24in. cylinders. The other is *No 22* built in 1923 and also has cylinders of 18in. diameter. Photographed at Waterside, Ayrshire, in 1966.

110 Andrew Barclay, 2335, built 1953

NCB *No 24* was the last side-tank delivered to the Dalmellington system and differed from previous examples in having sloping tops to the side tanks. When first used *No 24* was found to be a poor steamer and it was rebuilt with a Giesl ejector in place of the conventional chimney. This modification transformed the locomotive, both in appearance and performance, and it became a firm favourite with the train crews. It is seen at Waterside in 1966 having brought a load of coal along the 'branch' from the pits at Pennyvenie.

111 Andrew Barclay, 2135, built 1941

No 27 at Corby Steelworks in 1964 shunting a rake of slag ladles. The molten slag is tapped from the furnaces into the ladles and taken to the slag tip. One of the most arduous tasks in a steelworks, Barclay locomotives were often selected for this duty. *No 27* was supplied new to Wellingborough Ironworks and was transferred to Corby in the early 1960s.

112 Hudswell Clarke, 431, built 1895

Desborough Warren Ironstone Pits, Northamptonshire, boasted one of the oldest Hudswell Clarke locomotives in regular use in the mid-1960s. Originally built for the Sheepbridge Coal and Iron Company in 1895, it was rebuilt by them in 1928 and 1944, but still retains many original features, notably the sparse cab. Number 431 was photographed en route to the exchange sidings with a rake of privately owned iron-ore tippler wagons, based on British Railways' design.

113 Andrew Barclay, 1047, built 1905

Situated on the east side of the Kettering—Melton Mowbray line were the exchange sidings of Storefield Pits. These were owned by the South Durham Iron Company and on weekdays it was usual for up to three locomotives to be engaged on shunting duties between the sidings and the quarries. *No 11* is seen in 1965 hauling a rake of loaded iron-ore tippers up to the exchange sidings. It is a powerful locomotive, having 14in. x 22in. cylinders and 3ft. 5in. diameter wheels.

114 Peckett, 1289, built 1912

Another view of Storefield Pits shows *Cockspur* storming past the camera with a rake of loaded iron-ore tipplers for the exchange sidings. *Cockspur* is of Peckett class E having 15in. x 21in. cylinders. It was delivered new to Morris and Shaw's Birch Coppice Colliery, Warwickshire, and was sold to the Irchester Ironstone Company in 1924, spending time at Irchester before transfer to Storefield.

115 Kitson, 5477, built 1936

Barely a mile from Storefield were the Stewarts and Lloyds Glendon East Pits. One summer's day in 1965 *Caerphilly* was photographed in the quarry sidings. It is one of the Manning Wardle design locomotives supplied to Stewarts and Lloyds by Kitsons. Also of interest are the internal-user ironstone hopper-wagons.

116 Peckett, 1871, built 1934, metre gauge

A view in the Finedon quarries of Stewarts and Lloyds, Wellingborough, shows the method of operation employed. The skips are being loaded by a Ruston Bucyrus excavator and when full will be taken down to the tipping dock in the exchange sidings. The locomotive is *No 86,* one of two metre-gauge Pecketts supplied in 1934.

117 Peckett, 1867, built 1934

Between the wars, Peckett and Sons supplied a large number of locomotives to the Oxfordshire Ironstone Company at Banbury. One of their six-coupled products is seen in 1965 at Penhill Farm Sidings waiting to work a train of empty iron-ore tipplers to the loading dock. *Basic* is a class OX1 locomotive and was one of four Peckett 0-6-OST's owned by Oxfordshire Ironstone.

118 Robert Stephenson & Hawthorns, 7846, built 1955

One of Hams Hall power station's six-coupled side tanks is seen against a backscene of cooling towers, hauling coal from the BR exchange sidings. Built by Robert Stephenson and Hawthorns *No 13* was the last locomotive supplied to Hams Hall and may be compared with their first side tank (Fig 42) to see the final development of this design.

119 Peckett, 1990, built 1940

Two Peckett class W6 locomotives and a Bagnall saddle tank comprised the locomotive fleet at CEGB Ironbridge, Shropshire. They were all four-coupled locomotives and the Pecketts were powerful machines with 14in. x 22in. cylinders. *No 3* makes an impressive sight in 1968, while shunting at Ironbridge where it was delivered new.

120 Hudswell Clarke, 1730, built 1945

In 1968 Mexborough Power Station, Yorkshire, had only one locomotive, which was built by Hudswell Clarke. It was of their later design having a full-length, round-section, saddle tank and is seen here at the power station working from the exchange sidings. *No 2* had been returned to the makers for general overhaul and was resplendant in its lined black livery when photographed.

121 W. G. Bagnall, 2178, built 1921

This W. G. Bagnall locomotive is generally similar to *Worthington No 16* (Fig 9), but has a more austere appearance. The cab has also been fitted with a small coal bunker, but this is partly obscured by the shunter who is looking at the derailed Presflo wagon they had been shunting. One of eight locomotives delivered new to the Salsette Trombay Railway, Bombay, it was returned to England and sold to Alpha Cement, being transferred to Shipton on Cherwell, Oxfordshire, where it was photographed in 1967.

122 Peckett, 2112, built 1949

J.C.Staton and Company had an alabaster works at Tutbury, Staffordshire. The works was connected to the BR line by a half-mile long branch which left the exchange sidings and curved away at right angles, crossing the River Dove. The newer of their two Peckett locomotives is seen crossing the river bridge in 1964 with a rake of loaded hopper-wagons. It is a Class R4 engine having 12in. diameter cylinders and was supplied new to Truman's Brewery, Burton on Trent, who later sold it to Statons.

123 Robert Stephenson and Hawthorns, 7405, built 1948

Associated Portland Cement Manufacturers, Swanscombe, standardised exclusively on Hawthorn Leslie and Robert Stephenson and Hawthorns products from the late 1920s. They had seven locomotives, all four-coupled saddle tanks of which *No 7* illustrated was the last to be purchased. It was photographed in 1966 at Swanscombe Works hauling a train past the cliffs towards the quarry and was withdrawn four years later, together with the remaining steam locomotives.

124 Peckett, 1890, built 1936 and 1938, built 1937

It seems strange that a motor company should show a preference for steam locomotives, but the Ford Motor Company did just that. In 1966 they operated a total of four steam locomotives at Dagenham, all supplied by Peckett and Sons and two of them are seen working trains to and from the exchange sidings. They are both of Peckett Class B3 having 14in. x 22in. cylinders and 3ft. 7in. diameter wheels.

125 Hawthorn Leslie, 2491, built 1901

The British Celanese factory near Derby had three steam locomotives in latter years. The oldest was *Henry*, a veteran Hawthorn Leslie of 1901, seen taking tank wagons into the works in 1965. *Henry* was delivered new to Websters Brickworks Ltd, Coventry, and later acquired by Courtaulds Ltd, who transferred it to British Celanese in 1923.

126 Peckett, 1376, built 1915

One of British Aluminium Company's pretty Peckett saddle tanks is seen shunting in the works at Burntisland, Fife, in 1965. It is of Class M5 and has 10in. diameter cylinders. Built in 1915, it has an intermediate design of dome cover without the spring balances to the safety valves.

127 Hudswell Clarke, 1334, built 1918

I make no apology for returning to the Oxfordshire Ironstone Company's system for the final photograph. This picture of their first locomotive, *Sir Thomas,* heading away from the exchange sidings with a train of iron-ore wagons captures the atmosphere of industrial railways. Note that right-hand running is in operation, the opposite of normal British practice.

Some Preserved Industrial Locomotives

The locomotives listed were to be found at these locations in April 1975, but transfers may have taken place subsequently. An asterisk against the Fig number indicates that the illustration is similar to the locomotive listed, though not necessarily identical.

Name	Type	Cyls	Maker	No	Year Built	Fig
Bridgnorth, Shropshire, Severn Valley Railway Company.						
The Lady Armaghdale	0-6-OT	IC	HE	686	1898	
813	0-6-OST	IC	HC	555	1901	
Warwickshire	0-6-OST	IC	MW	2047	1926	75*
No.4	0-4-OST	OC	P	1738	1928	85*
Invicta	0-4-OST	OC	AB	2220	1946	19*
193	0-6-OST	IC	HE	3793	1953	60*
Carnforth, Lancashire, Lakeside Railway Company.						
John Howe	0-4-OST	OC	AB	1147	1908	33*
Jane Darbyshire	0-4-OST	OC	AB	1969	1929	19*
Gasbag	4 WVBT	VCG	S	8024	1929	
Faraday	0-4-OST	OC	HL	3793	1932	38*
Toby	0-4-OST	OC	AB	2134	1942	19*
Cranford No.2	0-6-OST	OC	WB	2668	1942	13
	0-4-OST	OC	AB	2230	1947	19*
6	0-6-0F	OC	WB	3019	1952	24*
David	0-4-OST	OC	AB	2333	1953	19*
	0-4-OST	OC	AB	2343	1943	19*
Chappel and Wakes, Colne, Essex, Stour Valley Railway Preservation Society.						
	0-4-OST	OC	AE	1875	1921	3*
Gunby	0-6-OST	IC	HE	2413	1941	60*
	0-6-OST	IC	RSH	7031	1941	115*
	0-6-OT	OC	RSH	7597	1949	118*
Jupiter	0-6-OST	IC	RSH	7671	1950	99*
90	0-6-OST	IC	HE	3790	1952	60*
Belvoir	0-6-OST	OC	AB	2350	1954	22*
40	0-6-OT	OC	RSH	7765	1954	100*
Chasewater, Staffordshire, Chasewater Light Railway Company.						
No.11	0-4-OST	OC	N	2937	1882	78*
	0-6-OST	OC	HC	431	1895	112
Asbestos	0-4-OST	OC	HL	2780	1909	37*
	0-4-OST	OC	AB	1223	1911	18*
Lion	0-4-OST	OC	P	1351	1914	85*
Didcot, Berkshire, Great Western Society						
Trojan	0-4-OST	OC	AE	1386	1897	1
Glenfield	0-4-OCT	OC	AB	880	1902	17*
	0-4-OF	OC	WB	2473	1932	24*
Bonnie Prince Charlie	0-4-OST	OC	RSH	7544	1949	36*
No.47	0-6-OST	OC	RSH	7800	1954	

Dilhorne, Staffordshire, Foxfield Light Railway.

	0-6-0ST	IC	MW	1317	1895	74*
Henry Cort	0-4-0ST	OC	P	933	1903	85*
	0-4-0ST	OC	HL	3581	1924	37*
	0-6-0ST	OC	AE	1919	1924	6*
Topham	0-6-0ST	OC	WB	2193	1927	13*
Lewisham	0-6-0ST	OC	WB	2221	1927	11*
	0-4-0F	OC	AB	1984	1930	24*
J. T. Daly	0-4-0ST	OC	WB	2450	1931	121*
Robert	0-6-0ST	OC	AE	2068	1933	6*
Hawarden	0-4-0ST	OC	WB	2623	1940	121*
11	0-4-0ST	OC	P	2081	1947	
	4 WVBT	VCG	S	9535	1952	93*
Wimblebury	0-6-0ST	IC	HE	3839	1956	60*

Embsay, Yorkshire, Yorkshire Dales Railway Society.

Banshee	0-6-0ST	OC	P	1870	1934	116*
No.2	0-4-0ST	OC	WB	2178	1921	121
Chemicals	0-4-0ST	OC	AB	1823	1924	18*
RS16	0-4-0ST	OC	AE	1908	1925	3*
No.5	0-6-0ST	OC	HC	1709	1939	47*
	0-4-0ST	OC	P	2085	1948	85*

Falkirk, Stirlingshire, Scottish Railway Preservation Society.

	0-4-0T	OC	AB	840	1899	
No.13	0-4-0ST	OC	N	2203	1876	78*
Ellesmere	0-4-0WT	OC	H(L)	244	1861	
No.1	0-6-0T	IC	NR	5710	1902	82*
Clydesmill No.3	0-4-0ST	OC	AB	1937	1928	19*
No.20	0-6-0T	IC	AB	2068	1939	25

Grosmont, Yorkshire, North Yorkshire Moors Railway Preservation Society.

No.3	0-4-0WT	OC	EB	37	1898	30*
29	0-6-2T	IC	K	4263	1904	69
5	0-6-2T	IC	RS	3377	1909	94*
No.2 Salmon	0-6-0ST	OC	AB	2139	1942	22*
Eustace Forth	0-4-0ST	OC	RSH	7063	1942	95*
No.1 Mirvale	0-4-0ST	OC	HC	1882	1955	120*
47	0-6-0ST	OC	RSH	7849	1955	97*

Haworth, Yorkshire, Keighley and Worth Valley Light Railway Limited.

Bellerophon	0-6-0WT	OC	HF		1874	
Sir Berkeley	0-6-0ST	IC	MW	1210	1891	74*
Lord Mayor	0-4-0ST	OC	HC	402	1893	
52	0-6-2T	IC	NR	5408	1899	94*
31 Hamburg	0-6-0T	IC	HC	679	1903	
67	0-6-0T	IC	HC	1369	1919	52*
75621	0-4-0ST	OC	P	1999	1941	85*
Fred	0-6-0ST	IC	RSH	7289	1945	60*
118 Brussels	0-6-0ST	IC	HC	1782	1945	60*
	0-4-0ST	OC	AB	2226	1946	19*
57	0-6-0ST	IC	RSH	7668	1950	99
62	0-6-0ST	IC	RSH	7673	1950	99*
63	0-6-0ST	IC	RSH	7761	1954	99*

Leeds, Yorkshire, Middleton Railway Trust.

Windle	0-4-OWT	OC	EB	53	1909	30*
Henry De Lacy II	0-4-OST	OC	HC	1309	1917	43*
54	4 WVBT	VCG	S	8837	1933	91*
No.6	0-4-OST	OC	HL	3860	1935	38*
Matthew Murray	0-4-OST	OC	WB	2702	1943	8*
	0-4-OST	OC	P	2003	1941	86*

Loughborough, Leicestershire, Main Line Steam Trust.

Littleton 5	0-6-OST	IC	MW	2018	1922	75*
Robert Nelson 4	0-6-OST	IC	HE	1800	1936	59*
39	0-6-OST	OC	RSH	6947	1938	96
Lamport No.3	0-6-OST	OC	WB	2670	1942	13*
	4 WVBT	VCG	S	9370	1947	93*
	4 WVBT	VCG	S	9373	1947	93*

Lytham St. Annes, Lancashire, Lytham Motive Power Museum.

Johnathan	0-4-OST	OC	HE	678	1898	53*
Hodbarrow No.6	0-4-OCT	IC	N	4004	1890	
Vulcan	0-4-OST	OC	VF	1828	1902	
No.20	0-4-OST	OC	NBL	18386	1902	80*
Penicuik	0-4-OST	OC	HL	3799	1935	38*
	0-4-OST	OC	HC	1661	1936	44*
Miranda	0-4-OST	OC	P	2087	1948	86*
Lytham No.1	0-4-OST	OC	P	2111	1949	85*
Susan	4 WVBT	VCG	S	9537	1952	93*

Quainton Road, Buckinghamshire, Quainton Railway Society.

Holy War	0-4-OST	OC	HE	779	1902	53*
Trym	0-4-OST	OC	HE	287	1883	
Swanscombe	0-4-OST	OC	AB	699	1891	33*
Sydenham	0-4-OWT	9	AP	3567	1895	
Punch Hull	0-4-OST	OC	AB	776	1896	33*
	0-4-OST	OC	MW	1795	1912	73
	0-4-OF	OC	AB	1477	1916	24*
Sir Thomas	0-6-OT	OC	HC	1334	1918	127
	0-4-OST	OC	AB	1865	1926	18*
Isebrook	4 WVBT	VCG	S	6515	1926	91
Hornpipe	0-4-OST	OC	P	1756	1928	85*
	0-4-OST	OC	HL	3717	1928	38*
Cunarder	0-6-OST	OC	HE	1690	1931	57*
	0-4-OST	OC	WB	2469	1932	121*
Tom Parry	0-4-OST	OC	AB	2015	1935	18*
	0-4-OT	OC	P	1900	1936	
Coventry No.1	0-6-OT	IC	NBL	24564	1939	82
	4 WVBT	VCG	S	9366	1945	92*
	0-4-OST	OC	HC	1742	1946	44*
	4 WVBT	VCG	S	9376	1947	93*
	0-4-OF	OC	AB	2243	1948	24*
	0-4-OST	OC	P	2104	1950	86*
	0-4-OST	OC	P	2105	1950	86*
	0-6-OST	OC	YE	2498	1951	96*
Juno	0-6-OST	IC	HE	3850	1958	60*

Shackerstone, Leicestershire, Shackerstone Railway Society.

The King	0-4-OST	OC	EB		1906	30
NCB II	0-4-OST	OC	HE	1493	1925	55
Lady Angela	0-4-OST	OC	P	1690	1926	86*
21	0-6-OST	OC	HL	3831	1938	39*
	0-4-OST	OC	P	2012	1941	84*
	0-6-OT	OC	RSH	7537	1949	42*
	0-6-OT	OC	RSH	7684	1951	42*
	0-4-OST	OC	P	2131	1951	86*

Sherringham, Norfolk, North Norfolk Railway.

Pony	0-4-OST	OC	HL	2918	1912	37*
Colwyn	0-6-OST	IC	K	5470	1933	115*
5	0-6-OST	OC	P	1970	1939	90*
Ring Haw	0-6-OST	IC	HE	1982	1940	59*
	0-6-OT	OC	AB	2107	1941	109*
12	0-6-OT	OC	RSH	7845	1955	118*

Tenterden, Kent, Kent and East Sussex Railway Society.

Minnie	0-6-OST	IC	FW	358	1878	
Hastings	0-6-OST	IC	HE	469	1888	58*
Arthur	0-6-OST	IC	MW	1601	1903	74
Met	0-4-OST	OC	HL	2800	1909	36
Dolobran	0-6-OST	IC	MW	1762	1910	75
Westminster	0-6-OST	OC	P	1378	1914	88
Charwelton	0-6-OST	IC	MW	1955	1917	75*
41	0-6-OST	IC	MW	2009	1921	75*
Marcia	0-4-OT	OC	P	1631	1923	
	0-4-OVBT	VC	S	6807	1928	
	4 WVBT	VCG	S	6994	1927	
	0-6-OST	IC	RSH	7667	1950	99*
	0-6-OST	IC	HE	3791	1952	60*
Errol Lonsdale	0-6-OST	IC	HE	3796	1952	60*
	0-6-OST	IC	HE	3800	1953	60*